T0187285

Cambridge Lower Secondary

Science

STAGE 7: STUDENT'S BOOK

Mark Levesley, Chris Meunier,
Fran Eardley, Gemma Young

Collins

William Collins' dream of knowledge for all began with the publication of his first book in 1819. A self-educated mill worker, he not only enriched millions of lives, but also founded a flourishing publishing house. Today, staying true to this spirit, Collins books are packed with inspiration, innovation and practical expertise. They place you at the centre of a world of possibility and give you exactly what you need to explore it.

Collins. Freedom to teach.

Published by Collins
An imprint of HarperCollins*Publishers*
The News Building
1 London Bridge Street
London
SE1 9GF

HarperCollins *Publishers*
1st Floor
Watermarque Building
Ringsend Road
Dublin 4
Ireland

Browse the complete Collins catalogue at
www.collins.co.uk

© HarperCollins*Publishers* Limited 2018

10 9 8 7 6

ISBN 978-0-00-825465-0

British Library Cataloguing in Publication Data
A catalogue record for this publication is available from the British Library.

Authors: Mark Levesley, Chris Meunier, Fran Eardley, Gemma Young
Development editors: Jane Glendening, Gillian Lindsey, Gina Walker
Team leaders: Mark Levesley, Peter Robinson, Aidan Gill
Commissioning project manager: Susan Lyons
Commissioning editors: Joanna Ramsay, Rachael Harrison
In-house editor: Natasha Paul
Copyeditor: Rebecca Ramsden
Proofreader: Elizabeth Barker
Technical checker: Mike Smith
Indexer: Jouve India Private Limited
Photo researcher: Alison Prior
Illustrator: Jouve India Private Limited
Cover designer: Gordon MacGlip
Cover artwork: Maria Herbert-Liew
Internal designer: Jouve India Private Limited
Typesetter: Jouve India Private Limited
Production controller: Tina Paul

Printed and bound in the UK using 100% Renewable Electricity at CPI Group (UK) Ltd

Contents

How to use this book

This book is designed to challenge you to go beyond the content you need to learn on your course. Have a go at the questions in dark green, blue and orange to challenge yourself, and read more about the scientific world in the discovering sections.

The outcomes show what you will learn

Chapter 1 . Topic 1

Characteristics of living things

Learning outcomes

- To identify the seven characteristics of living things
- To describe how scientists ask and answer questions
- To present data using tables, charts and graphs

This table helps remind you of what you know, and the scientific skills that you have. You will build on these as you study this topic

Starting point

You should know that...	You should be able to...
Plants and animals are living things	Use tables, bar charts and line graphs
Living things need certain things to survive – for example, plants need water	Use information from different places
Living things reproduce	

Working as a scientist

Scientists ask questions and think of ideas. They make observations and do experiments to test their ideas and answer their questions.

Before doing experiments, scientists make **predictions** to say what they think will happen. They also explain the ideas they used to make their predictions.

Observations and measurements are called **data**. Scientists use their data to show that their ideas are correct. They use their data as **evidence** for their ideas.

1 Write a list of steps a scientist takes to show that an idea is correct.

Organisms

Living things are called **organisms**. All organisms do certain things, called **life processes**, which keep them alive. The life processes also allow the different types of organism to continue to exist.

To show that something is living, we must collect evidence to show that it carries out all seven life processes:

- movement (moving part or all of themselves)
- reproduction (making new organisms like themselves)
- sensitivity (detecting changes)
- growth (getting bigger)

Key terms

data: numbers and words that can be organised to give information.

evidence: data or observations we use to support or oppose an idea.

life process: something that all living things do.

organism: living thing.

prediction: what you think will happen in an investigation.

1.1 *Ibn al Haytham (965–1039) lived in what is now Iraq. He was one of the first people to use experiments to answer questions, like scientists do today.*

Try the questions to check your understanding

You should learn the meanings of the key scientific terms in bold. You can find their meanings in the margin and in the glossary (near the end of the book)

Discovering microorganisms

Delft is a city in the Netherlands. In the 17th century, Antonie van Leeuwenhoek owned a fabric shop there. In order to check the threads in his fabrics, he made tiny magnifying glasses, which were very powerful. Using one of these, he was the first person to see organisms that are too small to observe with our eyes alone. Today, we know that these 'microorganisms' are made of only one cell.

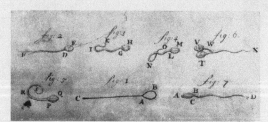

1.23 *Some of Antonie van Leeuwenhoek's original drawings made using his magnifying glasses.*

1 a) Why is cartilage tissue important at the ends of some bones?

b) What is cartilage tissue made of?

c) Explain why a bone is an organ.

d) What organ system do bones belong to?

2 Muscles contain many muscle cells.

a) Suggest a name for the tissue that these cells form.

b) Muscles also contain connective tissue, which is tough and strong. What part of a muscle do you think contains a lot of this tissue?

3 Suggest why scientists often think of blood as being a liquid organ.

Key facts:

✔ Cells form tissues, which form organs, which work together in organ systems in large organisms.

Check your skills progress:

I can present findings using words and drawings.

Discover more about where scientific ideas have come from and how they are used around the world now

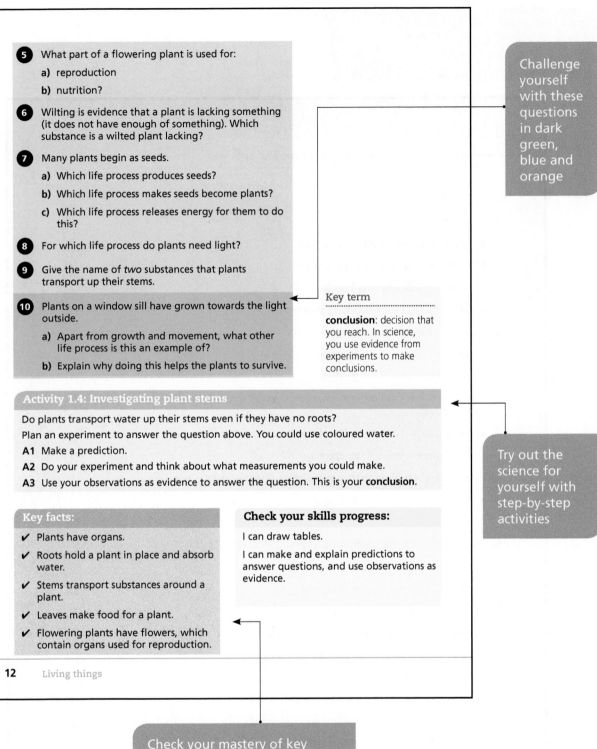

5. What part of a flowering plant is used for:

   a) reproduction

   b) nutrition?

6. Wilting is evidence that a plant is lacking something (it does not have enough of something). Which substance is a wilted plant lacking?

7. Many plants begin as seeds.

   a) Which life process produces seeds?

   b) Which life process makes seeds become plants?

   c) Which life process releases energy for them to do this?

8. For which life process do plants need light?

9. Give the name of *two* substances that plants transport up their stems.

10. Plants on a window sill have grown towards the light outside.

    a) Apart from growth and movement, what other life process is this an example of?

    b) Explain why doing this helps the plants to survive.

Challenge yourself with these questions in dark green, blue and orange

Key term

conclusion: decision that you reach. In science, you use evidence from experiments to make conclusions.

Activity 1.4: Investigating plant stems

Do plants transport water up their stems even if they have no roots?

Plan an experiment to answer the question above. You could use coloured water.

A1 Make a prediction.

A2 Do your experiment and think about what measurements you could make.

A3 Use your observations as evidence to answer the question. This is your **conclusion**.

Try out the science for yourself with step-by-step activities

Key facts:

✔ Plants have organs.

✔ Roots hold a plant in place and absorb water.

✔ Stems transport substances around a plant.

✔ Leaves make food for a plant.

✔ Flowering plants have flowers, which contain organs used for reproduction.

Check your skills progress:

I can draw tables.

I can make and explain predictions to answer questions, and use observations as evidence.

Check your mastery of key ideas and skills with this list

End of chapter review

Quick questions

1. In your body, a system is:

    a different tissues working together

    b different organs working together

    c different cells working together

    d different organisms working together [1]

2. A plant contains different organs, such as:

    a root hair   b water

    c palisade   d stem [1]

3. The life processes are movement, reproduction, growth, sensitivity, excretion, nutrition and:

    a respiration   b photosynthesis

    c replication   d stem [1]

4. To observe a specimen with a microscope, the specimen is put on a:

    a swing   b stage

    c lens   d slide [1]

5. The part of a cell that controls it is the:

    a nuclear   b newton

    c nucleus   d neutron [1]

6. One function of the skeletal system is protection.

    (a) Give the name of an organ protected by the skull. [1]

    (b) Give the name of an organ protected by the ribs. [1]

    (c) State *two* other functions of the skeletal system. [2]

7. (a) Make a drawing of an animal cell. [1]

    (b) Label the nucleus, cytoplasm and cell membrane. [3]

    (c) What is the function of the cell membrane? [1]

End of stage review

1. (a) (i) The table shows the parts of a cell. The functions are not in the correct order. Copy the table and put the functions of each part in the correct order. [1]

Part	Function
cell membrane	makes new substances
chloroplast	controls the cell
cytoplasm	makes food
nucleus	controls what enters and leaves the cell

(ii) Explain how you know that this is a plant cell. [1]

(b) The diagram shows some muscles in the leg.

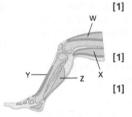

(i) Give the reason why many muscles are found in antagonistic pairs. [1]

(ii) State the letter of the muscle that contracts to point the toes out straight. [1]

(c) Four different trees are planted in the same area. Their heights are measured every year. The graph shows this data.

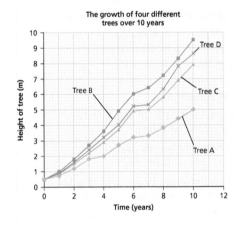

The growth of four different trees over 10 years

Glossary

Biology

acid rain: rain that is much more acidic than usual.

adaptation: feature of something that allows it to do a job (function) or allows it to survive.

amphibian: vertebrate with moist skin. It lays jelly-coated eggs in water.

animal kingdom: kingdom that contains organisms that are made of more than one cell and are able to move their bodies from place to place.

antagonistic pair: two muscles that pull a bone in opposite directions.

antiseptic: substance that kills microorganisms but is safe for us to put on our skins.

arachnid: arthropod with eight legs and a body in two sections.

arthropod: invertebrate with jointed legs and a body in sections.

bacterium: type of one-celled organism that is not a plant or animal or fungus. The plural is 'bacteria'.

ball and socket joint: joint where a ball-shaped piece of bone fits into a socket made by other bones.

bar chart: chart that shows data using columns. They are used to compare different sets of things.

biofuel: fuel made using plants or algae.

bird: vertebrate with feathers. It lays eggs with hard shells.

bladder: organ that stores urine.

blood: liquid organ that carries substances around the body.

blood vessels: tube-shaped organs that carry blood around the body.

bone: hard organ that supports or protects the body, or allows movement.

carnivore: animal that eats other animals.

cell: the smallest living part of an organism.

cell membrane: outer layer of a cell that controls what enters and leaves the cell.

cell wall: strong outer covering found in some cells (such as plant cells).

characteristic: feature of an organism.

chloroplast: green part of a cell that makes food using light.

circulatory system: group of organs that get blood around the body.

climate change: changes to weather patterns.

conclusion: decision that you reach. In science, you use evidence from experiments to make conclusions.

conifer: plant with needle-shaped leaves. It produces cones.

consumer: animal that eats other living things.

continuous variation: variation that can have any value within a range.

contract (muscle): when a muscle gets shorter and fatter it contracts.

cytoplasm: watery jelly where the cell makes new substances.

daily change: change in physical factors during the course of a day.

data: numbers and words that can be organised to give information.

decay: when materials break into smaller parts. Microorganisms often cause this.

decomposer: microorganism that causes decay.

deciduous: plant that loses its leaves during a certain season of the year.

deforestation: cutting down forests.

diagnosis: saying what disease someone has.

diaphragm: organ that helps with breathing.

digestive system: group of organs that digest food and get it into the blood.

discontinuous variation: variation that has a distinct set of options or categories.

You can look up definitions for key terms in the glossary

x

Biology

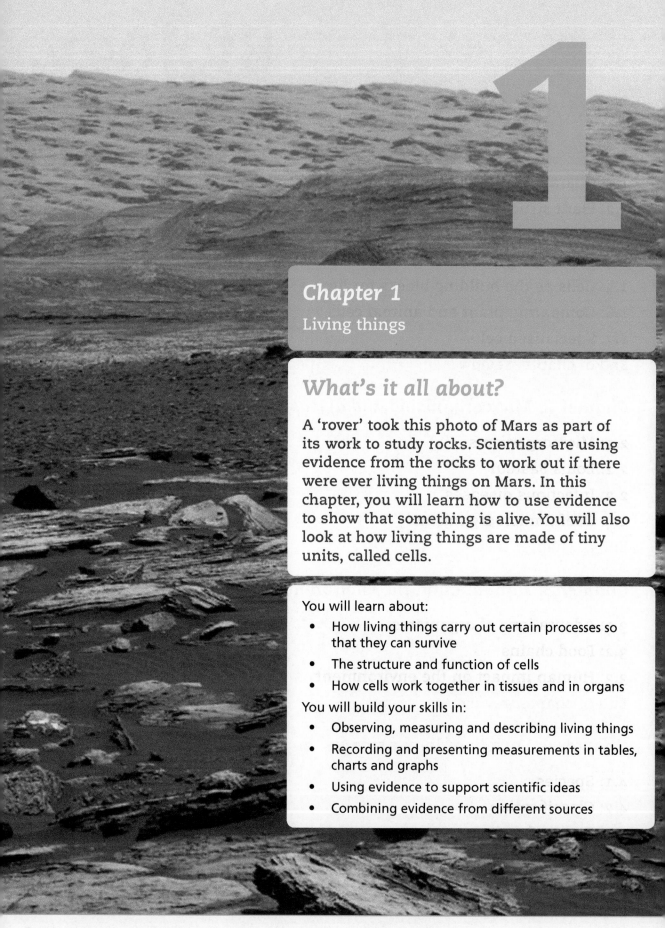

1

Chapter 1
Living things

What's it all about?

A 'rover' took this photo of Mars as part of its work to study rocks. Scientists are using evidence from the rocks to work out if there were ever living things on Mars. In this chapter, you will learn how to use evidence to show that something is alive. You will also look at how living things are made of tiny units, called cells.

You will learn about:
- How living things carry out certain processes so that they can survive
- The structure and function of cells
- How cells work together in tissues and in organs

You will build your skills in:
- Observing, measuring and describing living things
- Recording and presenting measurements in tables, charts and graphs
- Using evidence to support scientific ideas
- Combining evidence from different sources

Characteristics of living things

Learning outcomes
- To identify the seven characteristics of living things
- To describe how scientists ask and answer questions
- To present data using tables, charts and graphs

Starting point

You should know that...	You should be able to...
Plants and animals are living things	Use tables, bar charts and line graphs
Living things need certain things to survive – for example, plants need water	Use information from different places
Living things reproduce	

Working as a scientist

Scientists ask questions and think of ideas. They make observations and do experiments to test their ideas and answer their questions.

Before doing experiments, scientists make **predictions** to say what they think will happen. They also explain the ideas they used to make their predictions.

Observations and measurements are called **data**. Scientists use their data to show that their ideas are correct. They use their data as **evidence** for their ideas.

 1 Write a list of steps a scientist takes to show that an idea is correct.

Key terms

data: numbers and words that can be organised to give information.

evidence: data or observations we use to support or oppose an idea.

life process: something that all living things do.

organism: living thing.

prediction: what you think will happen in an investigation.

Organisms

Living things are called **organisms**. All organisms do certain things, called **life processes**, which keep them alive. The life processes also allow the different types of organism to continue to exist.

To show that something is living, we must collect evidence to show that it carries out all seven life processes:

- movement (moving part or all of themselves)
- reproduction (making new organisms like themselves)
- sensitivity (detecting changes)
- growth (getting bigger)

1.1 *Ibn al Haytham (965–1039) lived in what is now Iraq. He was one of the first people to use experiments to answer questions, like scientists do today.*

- respiration (providing energy)
- excretion (getting rid of wastes)
- nutrition (getting food).

To remember the life processes, make up a sentence using the first letter of each one. Or create a word or phrase using those letters. An example is: MRS GREN.

2 How many life processes are there?

3 Which life processes can you see in figure 1.2? Explain your choices.

1.2 *Barbary macaques live in North Africa.*

Movement

All organisms can move. Plants only move parts of themselves. Animals often move their whole bodies from place to place to find food and shelter, and to escape danger.

4 Why is movement essential for Barbary macaques?

Reproduction

Organisms **reproduce** to make new organisms like themselves. The new organisms are their **offspring**.

Scientists often count the numbers of organisms and their offspring in an area, especially for rare organisms. They show their data in tables and **bar charts**.

Area in Kazakhstan	Number of saiga antelope in 2015
Ural	51 700
Ustyurt	1 270
Betpak-Dala	242 500

Table 1.1 The numbers of rare saiga antelope in different parts of Kazakhstan in 2015.

Key terms

bar chart: a chart that shows data using columns. They are used to compare different sets of things.

offspring: new organism made when parents reproduce.

reproduce: when organisms have young (or offspring).

A bar chart makes the differences between values more obvious.

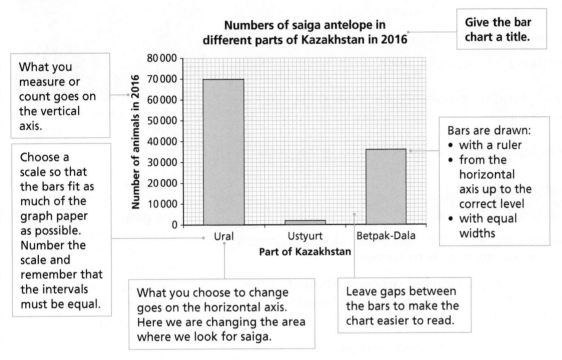

Give the bar chart a title.

What you measure or count goes on the vertical axis.

Choose a scale so that the bars fit as much of the graph paper as possible. Number the scale and remember that the intervals must be equal.

Bars are drawn:
• with a ruler
• from the horizontal axis up to the correct level
• with equal widths

What you choose to change goes on the horizontal axis. Here we are changing the area where we look for saiga.

Leave gaps between the bars to make the chart easier to read.

1.3 *Bar chart to show the numbers of saiga antelope in 2016.*

5 Present the data for saiga for 2015 in a bar chart.

6 Look at the bar chart in figure 1.3. How many saiga are there in the Ural region?

7 **a)** What has happened to the numbers of saiga in the Ural region between 2015 and 2016?

b) Which life process has allowed this to happen?

8 **a)** Saiga in one area got a disease. In which area did this happen? State the evidence you have used to work out your answer.

b) What may happen to the saiga in this area if there is not enough reproduction?

Sensitivity

Organisms sense things so they can react to changes in their surroundings. This is known as **sensitivity**. For example, many animals find food using their sense of smell and detect danger using their ears. Having good senses allows an organism to survive.

Humans have many senses, including taste, touch, hearing, smell, sight and balance.

Key term

sensitivity: how an organism detects changes in things inside and around it.

9 What parts of our bodies do we use to sense:

    a) the flavour of some ice cream

    b) someone shouting in the distance

    c) the feel of a piece of fabric?

10 Saiga are food for wolves. Explain how saiga use senses to survive.

11 Look at figure 1.4. What does a venus flytrap sense?

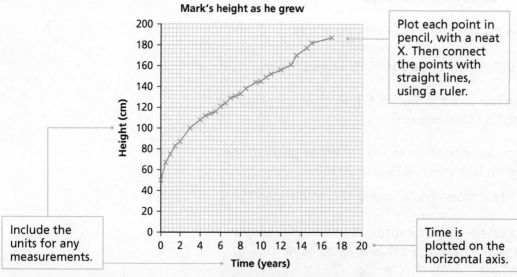

1.4 *Venus flytraps have special leaves that spring shut and trap insects.*

Growth

All organisms get bigger as they get older. Bigger organisms are stronger and so more likely to survive.

Some organisms stop growing but others continue to grow their whole lives. We show how something changes with time using a **line graph**.

Mark's height as he grew

Plot each point in pencil, with a neat X. Then connect the points with straight lines, using a ruler.

Include the units for any measurements.

Time is plotted on the horizontal axis.

Height (cm) / *Time (years)*

1.5 *Line graph to show the changes in Mark's height as he got older.*

12 Look at the graph in figure 1.5. How tall was Mark when he was 12 years old?

13 A lion cub is born with a mass of 0.6 kg. After 1 month it is 4.2 kg. After 2 months it is 9.0 kg and after 3 months it is 14.2 kg.

    a) Draw a table to show how the mass of the lion cub changes.

    b) Plot a line graph to show the growth of the lion cub.

14 State *one* difference between growth and reproduction.

15 How would you present the following data? Choose from bar chart or line graph and explain your choices.

   a) The masses of different breeds of dog when they are born.

   b) The change in saiga antelope numbers in one place over many years.

Respiration

Food contains a store of energy. This energy is released from the food by a chemical process called **respiration**. All organisms respire to release the energy they need.

Respiration in animals and plants usually needs oxygen, and produces waste carbon dioxide gas. You can detect carbon dioxide using **limewater**. If carbon dioxide is mixed with limewater, the clear and colourless liquid turns milky (or cloudy).

Key terms

limewater: clear and colourless liquid that turns milky when carbon dioxide is added.

respiration: chemical process that happens in all parts of an organism to release energy.

1.6 *If carbon dioxide is bubbled through limewater, the liquid turns milky.*

Respiration is not the same as breathing. When you breathe you move muscles to make your lungs bigger and smaller. Respiration is a chemical process that happens in every part of your body.

Activity 1.1: Investigating excretion of carbon dioxide

Do humans excrete carbon dioxide when they breathe out?

Plan an experiment to answer the question above.

A1 Make a prediction and explain your decision.

A2 Do your experiment and use your observations as evidence to answer the question.

Excretion

Organisms make waste substances, which can be poisonous and damage the organism. So, organisms **excrete** (get rid of) their wastes. Many animals excrete wastes in liquid **urine**.

 a) What life process produces carbon dioxide in humans?

b) How do we excrete carbon dioxide?

Uses of urine

The excreted substances in urine are surprisingly useful. In the past, French armies collected urine from their soldiers to make gunpowder. Urine was used in ancient Egypt to help add colour to fabrics. Today, a team of Korean scientists are trying to use urine to make fuel cells, which will produce clean and cheap electricity.

Activity 1.2: Investigating living things on Mars

Many scientists have wondered whether there is life on Mars. Some think that Mars is home to tiny organisms that excrete methane gas. These microorganisms are too small to see, so the idea is being tested by spacecraft looking at the gases on Mars. There is methane on Mars and the spacecraft are doing experiments to find out if this is (or was) gas being produced by living things, or if rocks are releasing it. One of the spacecraft involved is an Indian spacecraft called *Mangalyaan*.

A1 State the following:

a) the question that scientists have asked

b) the idea that some scientists have

c) what is being used to collect data.

Nutrition

Organisms need certain substances to survive. We say that they need **nutrition**.

Animals and humans get nutrition from their food. Plants make their own food (using energy from the Sun) but they also need small amounts of substances from the soil. Venus flytraps live in areas where there are not enough of these **nutrients** in the soil, and so they get these substances from insects.

1.7 *Different organisms get their nutrition in different ways.*

17 How is the way that cows get their nutrition different from the way that tigers get theirs?

18 Some driverless cars use petrol as a fuel. They avoid obstacles and so can carry people safely.

a) In what ways are these cars like organisms?

b) Why are they not living?

Activity 1.3: Investigating how plants respond to light

Do young plants grow towards light?

Plan an experiment to answer the question above. You could plant some seeds and use card to allow light to get to the seeds from only one direction.

A1 Make a prediction. Explain your prediction using ideas about life processes.

A2 Do your experiment and use your observations as evidence to answer the question.

Key facts:

✔ All organisms carry out seven life processes so that they can survive.

✔ The life processes are: movement, reproduction, sensitivity, growth, respiration, excretion, nutrition.

✔ Scientists collect data to use as evidence for their ideas and to answer their questions.

Check your skills progress:

I can draw tables, bar charts and line graphs, and know when to use them.

I can make and explain predictions to answer questions, and use data as evidence.

Plant structures

Learning outcomes
- To identify the main organs of a flowering plant
- To describe the functions of the main plant organs

Starting point

You should know that...	You should be able to...
Plants are living organisms and so carry out seven life processes	Present data using tables

Types of plants

There are many different types of plants. Some have **flowers** to reproduce and make seeds, and these are **flowering plants**. Plants that are not flowering plants use other ways of reproducing. For example, pine trees use cones to produce their seeds.

1 Why is a pine tree not a flowering plant?

2 Give the name of a flowering plant.

Key terms

flower: contains organs used in reproduction (to make seeds).

flowering plant: type of plant that produces flowers.

Uses of plants

Plants are very important in our lives. We use them for foods and flavourings, and to make fabrics, drinks, perfumes, dyes and building materials.

Some plant **roots** are used for food. Cassava roots and carrots are examples.

Thick tree **stems** (trunks) are used for building materials and to make paper. In Pakistan, chir pine trees are planted for this purpose.

Some plant **leaves** are used for dyes. For example, indigo is a blue substance made using the leaves of indigo plants. It is traditionally used to dye denim.

Some plant flowers are used for perfumes. For example, rose oil is extracted from the petals of rose flowers.

1.8 *Cassava being harvested in Vietnam.*

Key terms

leaf: plant organ that makes food for a plant.

root: plant organ that absorbs water from the ground, and holds the plant in place.

stem: plant organ that carries substances around a plant.

1.9 *Henna dye is extracted from the leaves of henna plants.*

3 Draw a table to show ten plants that you use. For each plant say which part of the plant you use and what you use it for. Make sure your table has clear headings, like this:

Name of plant	Part of plant that I use	What I use that part for

New uses for plant waste

Not all the parts of a plant are useful and this creates waste. Scientists try to invent ways of making useful things from these waste materials. For example, farmers traditionally burn banana plant stems or leave them to rot, but some people now make fabrics from them.

1.10 *Banana fabrics used for dresses in Sri Lanka.*

Plant organs

All parts of a plant need water and plants **wilt** (droop) if they do not have enough. Plants use roots to absorb (take) water from the ground. Roots are an example of an **organ** – a part of an organism with an important job (**function**). The drawing shows the main organs in flowering plants and what they do.

Key terms

function: another word for 'job'.

organ: part of an organism that has an important job (function).

wilt: when a plant droops because it does not have enough water.

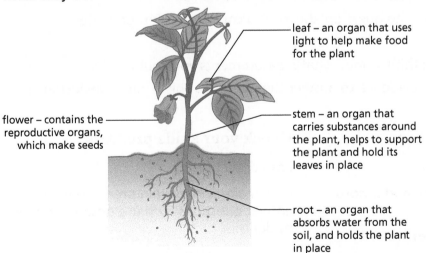

leaf – an organ that uses light to help make food for the plant

flower – contains the reproductive organs, which make seeds

stem – an organ that carries substances around the plant, helps to support the plant and hold its leaves in place

root – an organ that absorbs water from the soil, and holds the plant in place

1.11 *The functions of the main organs in a flowering plant.*

4 A pine tree is not a flowering plant. Draw a table to show the main organs in a pine tree and their functions.

5 What part of a flowering plant is used for:

a) reproduction

b) nutrition?

6 Wilting is evidence that a plant is lacking something (it does not have enough of something). Which substance is a wilted plant lacking?

7 Many plants begin as seeds.

a) Which life process produces seeds?

b) Which life process makes seeds become plants?

c) Which life process releases energy for them to do this?

8 For which life process do plants need light?

9 Give the name of *two* substances that plants transport up their stems.

10 Plants on a window sill have grown towards the light outside.

a) Apart from growth and movement, what other life process is this an example of?

b) Explain why doing this helps the plants to survive.

Key term

conclusion: decision that you reach. In science, you use evidence from experiments to make conclusions.

Activity 1.4: Investigating plant stems

Do plants transport water up their stems even if they have no roots?

Plan an experiment to answer the question above. You could use coloured water.

A1 Make a prediction.

A2 Do your experiment and think about what measurements you could make.

A3 Use your observations as evidence to answer the question. This is your **conclusion**.

Key facts:

✔ Plants have organs.

✔ Roots hold a plant in place and absorb water.

✔ Stems transport substances around a plant.

✔ Leaves make food for a plant.

✔ Flowering plants have flowers, which contain organs used for reproduction.

Check your skills progress:

I can draw tables.

I can make and explain predictions to answer questions, and use observations as evidence.

Skeleton, joints and muscles

Learning outcomes
- To identify some of the main bones and joints in the human body
- To describe how the skeleton is moved using muscles
- To explain why the muscles in joints are often found in pairs

Starting point

You should know that...	You should be able to...
Plants and animals are living organisms and so carry out seven life processes	Describe how scientists collect evidence to answer their questions
Plants and animals have organs	

Like plants, the bodies of animals and humans have important parts. Examples include **bones** and **muscles**, which are organs that help us to move.

 1 Which life process do bones and muscles help us with?

The skeletal system

We are each born with 270 bones but adults only have 206 bones! This is because as we grow our bones grow too, and some of them join with others.

 2 Which life process causes the number of bones in a human to decrease?

An **organ system** is a group of organs that work together. Your bones work together and form your **skeletal system** or **skeleton**. This has three main functions:

- support – to hold parts of your body in certain positions

- protection – to stop parts of your body being damaged

- movement – to let you move (using **joints**).

Key terms

bone: hard organ that supports or protects the body, or allows movement.

joint: place in your skeleton where bones meet.

muscle: organ that changes shape. Some muscles move bones.

organ system: group of organs working together.

skeletal system: all the bones in your body.

skeleton: another term for your skeletal system.

The drawing below shows some parts of the human skeletal system and their jobs.

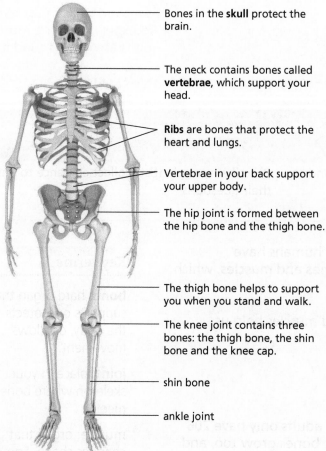

Bones in the **skull** protect the brain.

The neck contains bones called **vertebrae**, which support your head.

Ribs are bones that protect the heart and lungs.

Vertebrae in your back support your upper body.

The hip joint is formed between the hip bone and the thigh bone.

The thigh bone helps to support you when you stand and walk.

The knee joint contains three bones: the thigh bone, the shin bone and the knee cap.

shin bone

ankle joint

1.12 *The human skeletal system.*

Key terms

rib: bone that helps to protect your heart and lungs.

skull: a collection of bones that protect your brain.

vertebrae: the bones in your back. The singular is vertebra.

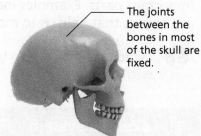

The joints between the bones in most of the skull are fixed.

1.13 *There are fixed joints in the skull.*

---

3 a) What are the *three* main jobs of the skeletal system?

b) For each job, give the name of a bone that helps with this job.

4 Give the names of the bones that form the knee joint.

Joints

Bones meet at places called joints. The joints between many bones in the skull are fixed and cannot move.

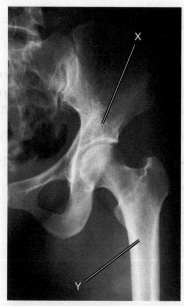

1.14 *Doctors use X-rays to study the bones inside our bodies.*

Other joints allow movement. An example is the hip joint. Here, the top of the thigh bone forms a ball that fits into a socket in the hip bone. This type of joint is called a **ball and socket joint**. It allows movement in many different directions.

5 a) Look at figure 1.14. What are the bones labelled X and Y?

b) What *type* of joint is this?

The elbow is a **hinge joint** (shown in figure 1.15). Hinge joints allow movement in two directions.

6 Your knee allows movement in two directions. What type of joint is it?

7 What type of joint is your shoulder? Give a reason for your answer.

Muscles and joints

Ligaments are cords that hold the bones in a joint in position. **Tendons** attach the muscles to the bones.

Muscles move the bones. A muscle pulls on a bone when it **contracts** (get shorter and fatter). When a muscle is not contracted we say it is **relaxed**.

Muscles only pull and cannot push. So bones are often moved by **antagonistic pairs** of muscles. One muscle in a pair pulls a bone in one direction. The other muscle pulls the bone in the opposite direction.

8 Look at the drawings of bones and muscles in the arm in figure 1.15.

a) Describe the change in shape when a muscle contracts.

b) Which muscle contracts to raise the lower arm?

c) Describe what happens to both muscles when the arm is lowered.

d) Explain why there is a pair of muscles to move the lower arm up and down.

Key terms

ball and socket joint: joint where a ball-shaped piece of bone fits into a socket made by other bones.

hinge joint: joint where two bones form a hinge.

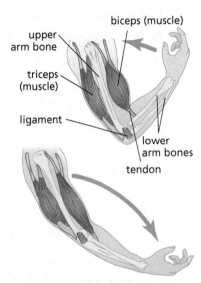

1.15 *The bones and two of the muscles in the arm.*

Key terms

antagonistic pair: two muscles that pull a bone in opposite directions.

contract (muscle): when a muscle gets shorter and fatter it contracts.

ligament: cord that attaches bones together.

relax (muscle): when a muscle stops contracting it relaxes.

tendon: cord that attaches muscles to bones.

Which arm muscles contract when you pull up or push down on a table edge?

Find a table (or a ledge) that will not move if you push down or pull up on it.

A1 Read the instructions below and make predictions. Then do the activity.

    A Turn the palm of your *right* hand to face upwards.

    B Keep your hand flat, and put just your fingers under the edge of the table.

    C Use your *left* hand to grip the front of your *right* arm, at the top.

    D Pull your *right*-hand fingers upwards, as if lifting the table. Record what you feel.

    E Now, grip the back of your *right* arm, at the top.

    F Pull your *right*-hand fingers upwards again.

    G Put your *right*-hand fingers on top of the table (palm facing up).

    H Repeat steps C–F above but this time push your fingers downwards.

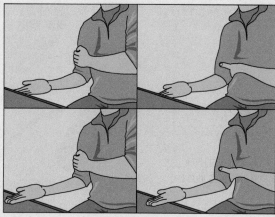

1.16

A2 Record your observations.

A3 Were your predictions correct?

A4 Explain your observations using scientific knowledge.

Walking robots

Scientists around the world are building machines that walk like humans. The scientists carefully study how the muscles and bones in human bodies work together. They work out the forces needed by different muscles to pull bones in different directions, and use this information to design ways of getting the robots to stand up and walk.

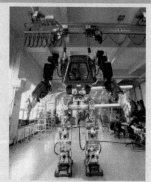

1.17 *This machine, designed in South Korea, could be used by scientists to explore volcanoes and other places where it is usually too dangerous for people to go.*

Injuries

Injuries may be painful and stop people moving easily. Common injuries include:

- fractures (broken bones)
- dislocations (when a bone comes out of place in a joint)
- pulled muscles (when a muscle or a tendon has been stretched too much)
- sprains (when a ligament has been stretched too much).

9 Use the evidence in the X-ray photo to:

a) identify which part of the skeletal system has been injured

b) state the kind of injury it is.

10 Give *one* similarity and *one* difference between a sprain and a pulled muscle.

11 Do you think there are more antagonistic pairs of muscles in a ball and socket joint or in a hinge joint? Explain your answer.

12 Sometimes the top of the thigh bone becomes worn and needs replacing. Explain *two* properties needed by a material used for a replacement.

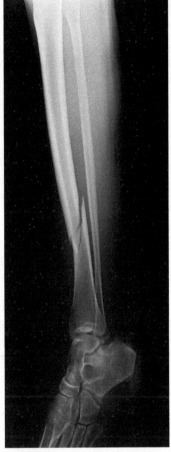

1.18 *Injuries to the skeletal system may stop it working properly.*

Activity 1.6: Investigating bones and joints

How do scientists investigate the bones and joints inside a living person?

X-rays are one way of seeing into a human. Find out about another method that doctors use to investigate bones, muscles and joints. Write *one* short paragraph to describe:

- what problems the method helps to find
- what happens to the person
- very briefly how the method works.

Key facts:

✔ The skeletal system protects and supports parts of the body, and allows the body to move.

✔ Muscles can only pull (when they contract), and so a bone in a joint needs antagonistic muscles to move it.

Check your skills progress:

I can make and explain predictions to answer questions, and use observations as evidence.

Human organs and organ systems

Learning outcomes

- To identify a range of human organs
- To describe the functions of the main human organs
- To describe how organs work together in different organ systems

Starting point

You should know that...	You should be able to...
Animals contain organs, such as bones and muscles	Describe how scientists collect evidence to answer their questions
Organs help organisms carry out the seven life processes	Present data using tables

Your body contains many muscles and bones but there are other organs that you only have one or two of. For example, you have two lungs, one heart and one brain. Figure 1.19 shows some other main organs in your body, and their functions.

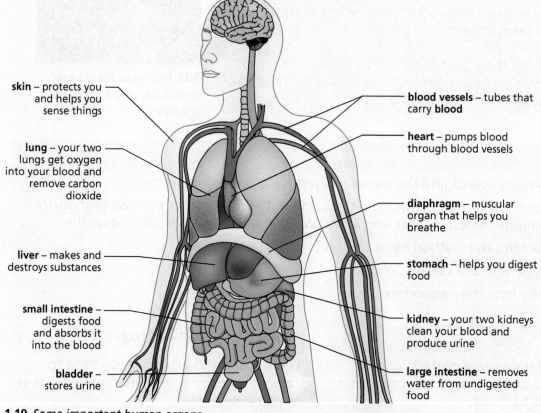

skin – protects you and helps you sense things

lung – your two lungs get oxygen into your blood and remove carbon dioxide

liver – makes and destroys substances

small intestine – digests food and absorbs it into the blood

bladder – stores urine

blood vessels – tubes that carry **blood**

heart – pumps blood through blood vessels

diaphragm – muscular organ that helps you breathe

stomach – helps you digest food

kidney – your two kidneys clean your blood and produce urine

large intestine – removes water from undigested food

1.19 *Some important human organs.*

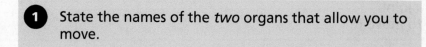

1 State the names of the *two* organs that allow you to move.

2 Give an example of an organ that people normally have two of.

3 Which life process do the following organs help with?

a) lungs   b) kidneys

c) small intestine   d) skin

4 Which human organ(s):

a) makes and destroys substances

b) transports blood all around the body?

Organ transplants

The earliest organ transplants were 'skin grafts', which were carried out 3000 years ago in a region that is now India. In a skin graft, a doctor takes a layer of skin from one part of a person's body and puts it on a damaged area of skin on the same person. Doctors around the world do hundreds of thousands of skin grafts every year, often to treat people with burns.

It is much more difficult to transplant organs between different people. The first successful organ transplant from one person to another was a kidney transplant operation in the USA in 1954.

Our bodies attack and destroy organs that do not belong to us. Scientists around the world continue to develop ways to stop this happening, and so allow more organ transplants. Thanks to this research doctors now transplant many different organs, including hearts, lungs, livers and even whole faces.

1.20 *A French woman called Isabelle Dinoire was the first person to receive a face transplant, in 2005, after an attack by a dog.*

5 Discover when doctors first *successfully* transplanted these organs between people: liver, heart, both lungs. Present your work as a table, showing the organ, the year of the first successful transplant and the country.

Key terms

bladder: organ that stores urine.

blood: liquid organ that carries substances around the body.

blood vessels: tube-shaped organs that carry blood around the body.

diaphragm: organ that helps with breathing.

heart: organ that pumps blood through blood vessels.

kidneys: organs that remove wastes from the blood to produce urine.

large intestine: organ that absorbs water from undigested food.

liver: organ that makes and destroys substances.

lungs: organs that get oxygen into the blood and remove carbon dioxide.

skin: organ that protects the body and helps it sense things.

small intestine: organ that digests food and absorbs it into the blood.

stomach: organ that helps to digest food.

Which part of your skin is most sensitive?

A1 Predict which part of your skin is most sensitive to touch. Choose from:

- tip of middle finger
- back of middle finger
- back of hand
- palm.

A Bend a piece of wire into a U-shape, with the ends 5 mm apart.

B Work with a partner. One of you shuts your eyes (or puts on a blindfold). The other person then *gently* presses the ends of the wire onto the skin in the different places in the list above. Each time you press the wire, the person with their eyes shut says whether they feel one point or two points.

C Repeat step B with a 1 mm gap between the ends of the wire.

A2 Record your observations in a table.

A3 Was your prediction correct?

A4 Explain how your results tell you which part of the skin on your hand is most sensitive. This is your conclusion.

Organ systems

Organs in your body work together in organ systems. You have already met the skeletal system. Here are some organ systems:

- **circulatory system** – gets blood to all parts your body
- **nervous system** – controls your body
- **respiratory system** – gets oxygen into your blood and removes carbon dioxide
- **digestive system** – digests your food and absorbs it into your blood.

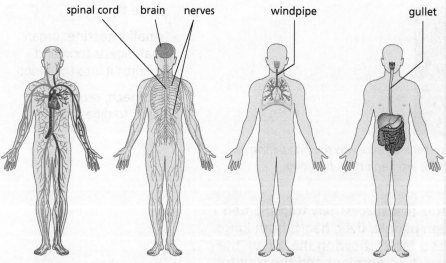

spinal cord   brain   nerves   windpipe   gullet

circulatory system   nervous system   respiratory system   digestive system

1.21 *Some human organ systems.*

Key terms

circulatory system: group of organs that get blood around the body.

digestive system: group of organs that digest food and get it into the blood.

nervous system: group of organs that control the body.

respiratory system: group of organs that get oxygen into the blood and remove carbon dioxide. Also called the breathing system.

6 Draw a table to show *five* different organ systems in the body, their functions and the organs that they contain.

7 Another human organ system is the excretory system.

 a) What is the function of this system?

 b) Give the names of *two* organs in this system.

Key facts:

✔ Main organs in the body include the skin, lungs, blood vessels, heart, diaphragm, liver, stomach, small intestine, large intestine, kidneys, bladder.

✔ Many organs work together in organ systems, such as the circulatory system, the nervous system, the respiratory (breathing) system and the digestive system.

Check your skills progress:

I can make and explain predictions to answer questions, and use observations as evidence.

I can present data as tables.

Cells as the building blocks for life

Learning outcomes
- To describe how organs are made up of tissues
- To describe how tissues are made up of cells
- To describe organisms in terms of cells, tissues, organs and organ systems

Starting point

You should know that...	You should be able to...
Animals contain many different organs, which work together in organ systems	Make and record observations

Figure 1.22 shows a thigh bone. Not all its parts look the same. These different areas are its **tissues**. All organs are made of different tissues.

Tissues are made of smaller parts called **cells**. Cells are the smallest living parts of organisms, and all the cells in a tissue are the same.

In a large organism, such as a plant or an animal:

- a group of cells of the same type forms a tissue

- a group of tissues working together forms an organ

- a group of organs working together forms an organ system.

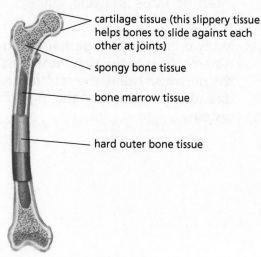

cartilage tissue (this slippery tissue helps bones to slide against each other at joints)

spongy bone tissue

bone marrow tissue

hard outer bone tissue

1.22 *A human thigh bone.*

Activity 1.8: Investigating chicken wings

Can you identify some different tissues and organs in a chicken wing?

Have a look at a chicken wing before it is cooked. Try to find the following parts:

- two different organs

- tendons

- cartilage (it is a different colour to the rest of the bone).

A1 Write a short report to describe what you have found and include a drawing. Do not use more than 200 words.

Key terms

cell: the smallest living part of an organism.

tissue: group of cells of the same type.

Discovering microorganisms

Delft is a city in the Netherlands. In the 17th century, Antonie van Leeuwenhoek owned a fabric shop there. In order to check the threads in his fabrics, he made tiny magnifying glasses, which were very powerful. Using one of these, he was the first person to see organisms that are too small to observe with our eyes alone. Today, we know that these 'microorganisms' are made of only one cell.

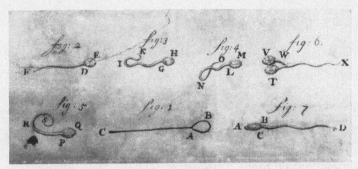

1.23 *Some of Antonie van Leeuwenhoek's original drawings made using his magnifying glasses.*

1
  a) Why is cartilage tissue important at the ends of some bones?

  b) What is cartilage tissue made of?

  c) Explain why a bone is an organ.

  d) What organ system do bones belong to?

2 Muscles contain many muscle cells.

  a) Suggest a name for the tissue that these cells form.

  b) Muscles also contain connective tissue, which is tough and strong. What part of a muscle do you think contains a lot of this tissue?

3 Suggest why scientists often think of blood as being a liquid organ.

Key facts:

✔ Cells form tissues, which form organs, which work together in organ systems in large organisms.

Check your skills progress:

I can present findings using words and drawings.

Comparing plant and animal cells

Learning outcomes
- To identify the main parts of cells
- To compare and contrast animal and plant cells
- To describe how to use a microscope correctly

Starting point

You should know that...	You should be able to...
Animals contain many different organs	Plan to use a magnifying glass to magnify things
Organs are made of different tissues	
A tissue is made of one type of cell	

In order to see cells we need to make them appear larger. We **magnify** them using **microscopes**.

Using microscopes, we know that cells have three main parts:

- **nucleus** – controls what the cell does

- **cell membrane** – controls what enters and leaves the cell

- **cytoplasm** – a watery jelly, where the cell makes new substances.

1 Which part of a cell controls the whole cell?

2 How many times bigger does a cell appear if it is magnified 400 times?

Figure 1.24 shows an animal cell. Plant cells usually have a more boxy shape than animal cells, and contain three other parts:

- **chloroplasts** – make food for the plant

- **cell wall** – strong outer covering that helps to support and protect the cell

- **vacuole** – storage space, which also helps the cell to keep its shape by pushing the cytoplasm against the cell wall.

3 Draw a table to show the names of *six* common cell parts and whether they are usually found in animal cells, plant cells or both.

4 Explain why plant cells are often green but animal cells are not.

Key terms

magnify: to make something appear bigger.

microscope: piece of equipment that magnifies very small things.

nucleus: control centre of a cell.

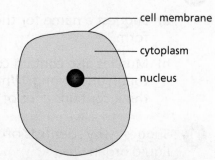

1.24 *An animal cell, magnified about 500 times. This means that it has been drawn 500 times larger than real life.*

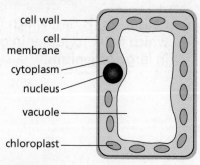

1.25 *A plant cell, magnified about 800 times.*

Activity 1.9: Investigating cells 1

How can you make a **model** of a plant or animal cell?

Design a model of a plant or animal cell. You could also try to make your model.

- Think about what you could use for each part. Maybe you could use a plastic bag for a cell membrane and a stone for a nucleus.

A1 Describe *one* way in which your model is good at helping you to understand cells.

A2 Describe *one* way in which your model is not so good at showing what a cell is like.

Microscopes

Robert Hooke used a microscope to discover cells in 1665. He was studying bark from cork oak trees. People grow these trees in Morocco and other Mediterranean countries to make corks. When magnified, he saw that the bark was made of tiny 'boxes'. They reminded him of the rooms in a monastery or prison, and so he called them 'cells'.

Figure 1.26 shows a modern microscope. The thing you want to magnify is the **specimen**. You place the specimen on the **stage**, usually held in place on a small sheet of glass called a **slide**. Light shines up from a light source or mirror, through the specimen and into an **objective lens**. You look through another lens – the **eyepiece lens**. The microscope needs light to work, and so we call it a light microscope.

5 a) What is a lens above the stage called?

b) How many of this type of lens are in the microscope in figure 1.26?

6 Explain why the stage has a hole in it, just above the light source.

Inventing the modern microscope

Symbols called hieroglyphs give us evidence that the Ancient Egyptians used glass lenses to magnify things 7000 years ago. A thousand years ago, the Arabic scientist Ibn al-Haytham (see Topic 1.1) wrote a series of important books about lenses and his ideas about how they worked.

In the 1590s, two Dutch spectacle makers (Zacharias and Hans Jansen) worked out that you could get a greater **magnification** by using two lenses together in a tube. They had invented the modern microscope.

Key terms

cell membrane: outer layer of a cell that controls what enters and leaves the cell.

cell wall: strong outer covering found in some cells (such as plant cells).

chloroplast: green part of a cell that makes food using light.

cytoplasm: watery jelly where the cell makes new substances.

model: simple way of showing or explaining a complicated object or idea.

slide: small sheet of glass on which you place a thin specimen.

specimen: the thing you examine using a microscope.

stage: flat surface on a light microscope where you put a slide.

vacuole: storage space inside some cells (such as plant cells).

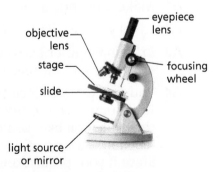

1.26 *The parts of a light microscope.*

Using a microscope

The steps below describe how to use a light microscope.

A Turn the objective lenses to put the smallest one over the hole in the stage. This lens gives the lowest magnification.

B Turn the **focusing wheel** to make the distance between the stage and the objective lens as small as possible.

C Adjust the light source so that light shines through the hole in the stage. Do *not* point the mirror at the Sun. This could damage your eyes permanently.

D Put the slide into the clips on the stage.

E Look through the eyepiece lens.

F Turn the focusing wheel until the image is in focus (clear and sharp).

G Move a larger objective lens over the specimen. *Very* slowly turn the focusing wheel until the image is in focus. If you turn the focusing wheel too much you could break the slide and damage the objective lens.

7 When using a microscope, describe *one* way in which you would:

   a) remain safe

   b) avoid damaging the microscope.

Activity 1.10: Investigating cells 2

What do cells look like under a microscope?

Figures 1.27 and 1.28 show cells viewed under a microscope.

A1 Make drawings of *one* cell from each image.

A2 Label your drawings to show the parts.

A3 Explain whether each cell in your drawings shows an animal or a plant cell. This is your conclusion.

A4 You may get a chance to look at slides. For each slide, find out where the cells are from and predict what they will look like. Use the microscope to make a labelled drawing of each type of cell that you see. State if your predictions were correct.

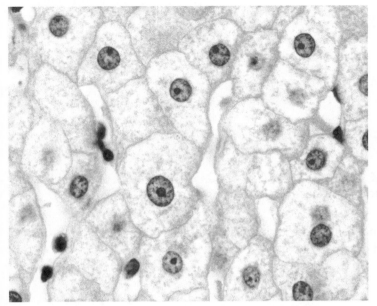

1.27 *Cells seen under a microscope. The cells have been stained with a purple dye to make different parts show up better.*

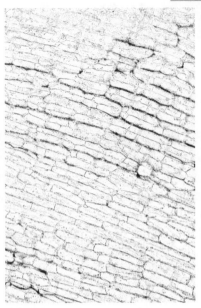

1.28 *More cells seen under a microscope.*

8 You need to find out if a specimen of tissue is from a plant or an animal.

    a) What piece of equipment will you use?

    b) How will you decide where the cells are from?

9 Draw a table to show the different parts of a microscope and what they do.

10 Explain why a specimen needs to be thin.

11 A heart muscle cell has a long, cylinder shape. It is 0.1 mm long and 0.02 mm wide. How long and how wide would the cell appear if magnified 500 times?

Key facts:

✔ Animal cells have a nucleus, cytoplasm and a cell membrane.

✔ Plant cells also have a vacuole and a cell wall, and often have chloroplasts.

✔ A light microscope is used to examine slides containing specimens.

Check your skills progress:

I know when to use a light microscope.

I can stay safe when using a light microscope.

I can use a light microscope without damaging it.

Specialised cells

Starting point

You should know that...	You should be able to...
An animal cell has a cell membrane, a nucleus and cytoplasm	Describe how to use a microscope
A plant cell has the same parts as an animal cell but also has a cell wall and a vacuole and often contains chloroplasts	Present data using tables

Plant cells in root tissues do not contain chloroplasts. That is because chloroplasts need light and there is no light underground.

Leaves contain a layer of **palisade cells**. These cells have many chloroplasts that use light to make food. A cell with a certain feature to do a certain job is 'adapted for its function'. The cell has an **adaptation**.

> **1** Explain why cells from an onion do not contain chloroplasts.
>
> **2** a) What is the function of palisade cells?
>
>   b) What adaptation do palisade cells have for this function?

On the outside of many roots, is a layer of root hair tissue. It is made of **root hair cells**, which have bits sticking out them that look a bit like hairs. A 'root hair' gives a cell a lot of **surface area**, which helps it absorb water quickly.

Key terms

adaptation: feature of something that allows it to do a job (function) or allows it to survive.

palisade cell: cell found in plant leaves, which contains many chloroplasts.

root hair cell: plant cell found in roots that is adapted for taking in water quickly.

surface area: the area of a surface, measured in squared units such as square centimetres (cm^2).

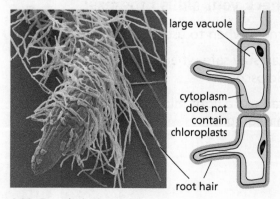

large vacuole

cytoplasm does not contain chloroplasts

root hair

1.29 *Root hair tissue.*

3 Identify *two* ways that a root hair cell is different from a leaf cell.

4 Explain how a root hair cell is adapted to its function.

Activity 1.11: Investigating root hair cells

How can you use a sponge to model how a root hair cell works?

You will need a cuboid-shaped sponge, a ruler, a watch or clock with a second hand, a cup or beaker, and a flat dish or tray of water.

A To work out the area of a rectangle you multiply its width by its length. Your sponge has six surfaces – each is a rectangle. For each surface, measure its length and its width in centimetres. Then calculate its surface area in square centimetres (cm^2) and write it down.

B Put a layer of water in your tray or dish and mark the top of the water level.

C Place one side of your sponge in the water and time 5 seconds.

D Pick up the sponge and squeeze *all* of its water into the cup.

E Mark the water level in the cup.

F Add more water to your tray or dish to the same level it was in step B.

G Repeat steps C–F to compare the **volume** of water soaked up by each surface of the sponge.

A1 Show your results in a table.

A2 Which surface soaked up the greatest volume of water in 5 seconds?

A3 Write out the order in which the surfaces soaked up the water. Start with the surface that soaked up the least.

A4 Write a conclusion for your experiment by completing this sentence:

The greater the _____ of the surface, the _____.

A5 Use your conclusion to explain why root hair cells with a large surface area are helpful for a plant.

Key term

volume: how much space a substance takes up. Measured in cm^3 or litres. Also called 'capacity'.

Specialised animal cells

A specialised cell is a cell that is adapted for a certain function. Palisade cells and root hair cells are **specialised cells**.

Animals have specialised cells too. For example, muscle cells contain special fibres that allow them to contract and relax. Some other examples are shown in figure 1.30.

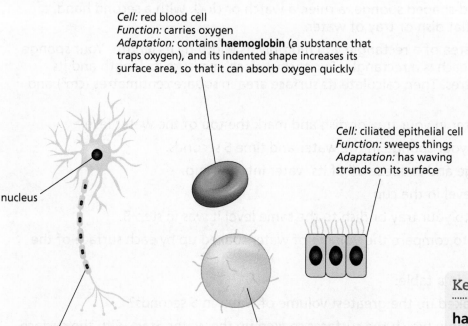

Cell: red blood cell
Function: carries oxygen
Adaptation: contains **haemoglobin** (a substance that traps oxygen), and its indented shape increases its surface area, so that it can absorb oxygen quickly

Cell: ciliated epithelial cell
Function: sweeps things
Adaptation: has waving strands on its surface

nucleus

A nerve cell carries signals from one part of the body to another. It is very long, to help carry these signals quickly.

A white blood cell has a very flexible shape allowing it to squeeze into all the different parts of the body. It finds and destroys invading cells inside the body.

1.30 *Some specialised animal cells.*

Key terms

haemoglobin: substance that traps oxygen.

specialised cell: cell with adaptations for a certain job.

Stem cells

'Stems cells' are cells that are able to develop into many different types of specialised cells. Scientists are developing ways of using stem cells to repair damaged tissues and organs. They hope that people will be injected with stem cells that then develop into the specialised cells needed to repair damage. For example, some people cannot move parts of their bodies because they have damaged nerves in their nervous systems. Scientists hope that, if they can inject stem cells into the damaged parts, the stem cells will develop into new nerve cells to repair the problem.

1.31 *Many scientists in South Korea are working on stem cell treatments.*

5 a) What is meant by the term 'specialised cell'?

b) Give the name of *one* specialised cell found in animals.

c) Give the name of *one* specialised cell found in plants.

6 Look at figure 1.30 showing some specialised animal cells. Rewrite the labels for the nerve cell and the white blood cell to match the labels of the other two cells (clearly showing the cell's name, its function and its adaptation).

7 Why is a 'root hair' not really like a hair?

Key facts:

✔ Specialised cells have adaptations so that they can do certain jobs.

✔ Specialised plant cells include palisade cells and root hair cells.

✔ Specialised animal cells include nerve cells, blood cells and muscle cells.

Check your skills progress:

I can calculate the area of a rectangle.

I can use evidence to make conclusions from results.

I can present information as a table.

Quick questions

1. In your body, a system is:

   a different tissues working together

   b different organs working together

   c different cells working together

   d different organisms working together   [1]

2. A plant contains different organs, such as:

   a root hair   b water

   c palisade   d stem   [1]

3. The life processes are movement, reproduction, growth, sensitivity, excretion, nutrition and:

   a respiration   b photosynthesis

   c replication   d stem   [1]

4. To observe a specimen with a microscope, the specimen is put on a:

   a swing   b stage

   c lens   d slide   [1]

5. The part of a cell that controls it is the:

   a nuclear   b newton

   c nucleus   d neutron   [1]

6. One function of the skeletal system is protection.

   (a) Give the name of an organ protected by the skull.   [1]

   (b) Give the name of an organ protected by the ribs.   [1]

   (c) State *two* other functions of the skeletal system.   [2]

7. (a) Make a drawing of an animal cell.   [1]

   (b) Label the nucleus, cytoplasm and cell membrane.   [3]

   (c) What is the function of the cell membrane?   [1]

8. Copy and complete this table to show if these parts of cells are found in animal cells, plant cells or both. Complete your table with ticks (✓). One row has been done for you.

cell part	animal cell	plant cell
cell membrane	✓	✓
cell wall		
chloroplast		
cytoplasm		
large vacuole		
nucleus		

[5]

9. Give the name of *one* organ in each of these organ systems:

(a) digestive system

(b) respiratory system

(c) circulatory system. [3]

Connect your understanding

10. The diagram shows some muscles and bones around the knee in a human leg.

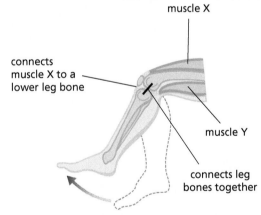

muscle X

connects muscle X to a lower leg bone

muscle Y

connects leg bones together

1.32 *The muscles and bones around the knee joint.*

(a) Give the name of the part that connects a muscle to a bone. [1]

(b) Give the name of the part that connects bones together. [1]

(c) Which muscle has to contract so that the lower leg moves in the direction shown by the arrow? [1]

(d) What happens to the other muscle during this movement? [1]

(e) What are pairs of muscles like this called? [1]

(f) Explain why bones must be moved by two muscles, rather than just one. [1]

(g) Which life process do muscles help with? [1]

(h) Give the name of the organ system formed by all the bones in the body. [1]

(i) What type of joint is the knee joint? [1]

11. The respiratory system gets oxygen into the blood.

(a) For which life process is oxygen needed? [1]

(b) Which cells carry oxygen around the body in the blood? [1]

(c) Explain *one* way in which these cells are adapted for their function. [2]

(d) What piece of equipment would you use to look at these cells in detail? [1]

12. (a) State the function of leaves in a plant. [1]

(b) There are many palisade cells in a leaf. Give the name of the tissue that they form. [1]

(c) Explain how palisade cells are adapted for their function. [2]

(d) At what times during a day do these cells perform this function? Give a reason for your answer. [2]

(e) Which life process do leaves help with? [1]

13. The drawing shows a type of specialised cell called a nerve cell.

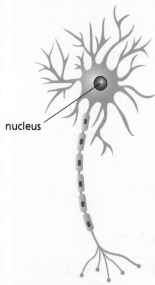

nucleus

1.33 *A nerve cell.*

(a) Is this a plant or an animal cell? Explain how you know. [2]

(b) Explain how this cell is adapted to its function. [2]

(c) Which life process does the cell use to get a supply of energy? [1]

14. The number of chloroplasts in some different plant cells was counted. The results are shown on the bar chart.

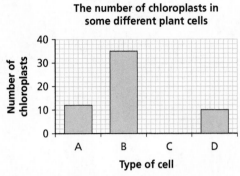

The number of chloroplasts in some different plant cells

1.34 *Bar chart.*

(a) How many chloroplasts were in the type A cell? [1]

(b) Which cell probably comes from a root? Explain your reasoning. [2]

(c) Type A and B cells are found in leaves. Suggest a name for cell type B. [1]

15. (a) When using a microscope, the specimen is very thin. Explain why. [1]

(b) A specimen is usually put on a small piece of glass. What is this called? [1]

(c) Which part of the microscope do you turn to make a clear and sharp image? [1]

(d) Explain *one* safety rule that you need to use when working with a microscope. [2]

Challenge questions

16. Yeasts are tiny organisms made of only one cell. In an experiment, yeast cells were grown in tubes containing sugar dissolved in water. They made a gas. This was bubbled through limewater, which slowly became milky.

(a) What was the gas? Give a reason for your answer. [2]

(b) What process produces this gas? [1]

17. The drawing shows a specialised cell from a plant. Suggest the function of this cell. Explain your reasoning. [2]

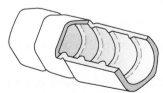

1.35 *A specialised plant cell.*

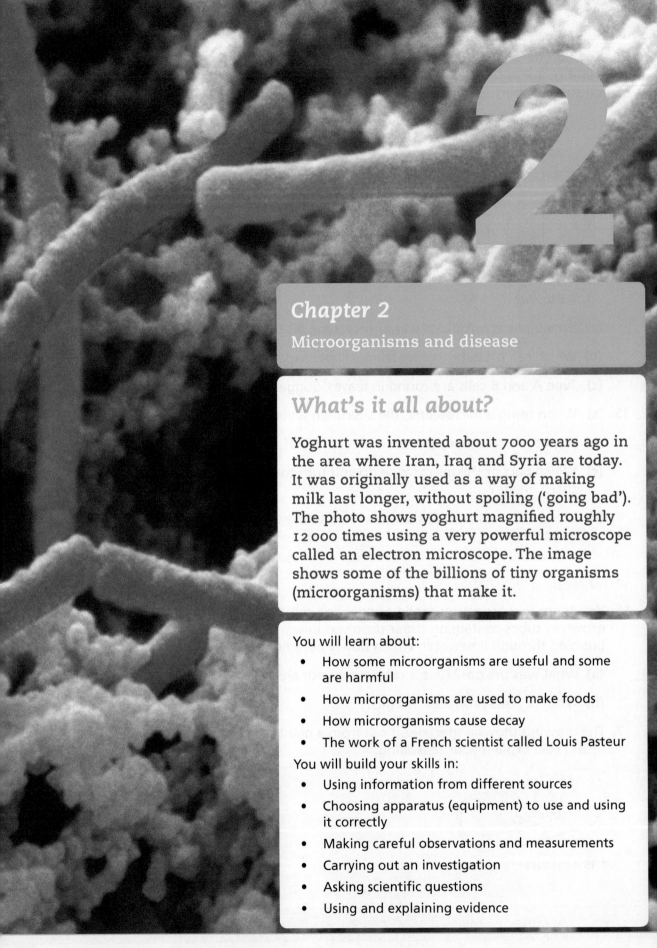

Chapter 2

Microorganisms and disease

What's it all about?

Yoghurt was invented about 7000 years ago in the area where Iran, Iraq and Syria are today. It was originally used as a way of making milk last longer, without spoiling ('going bad'). The photo shows yoghurt magnified roughly 12 000 times using a very powerful microscope called an electron microscope. The image shows some of the billions of tiny organisms (microorganisms) that make it.

You will learn about:
- How some microorganisms are useful and some are harmful
- How microorganisms are used to make foods
- How microorganisms cause decay
- The work of a French scientist called Louis Pasteur

You will build your skills in:
- Using information from different sources
- Choosing apparatus (equipment) to use and using it correctly
- Making careful observations and measurements
- Carrying out an investigation
- Asking scientific questions
- Using and explaining evidence

Microorganisms

Starting point

You should know that...	You should be able to...
An organism can carry out seven life processes, and what these life processes are	Explain how scientists think up ideas, make predictions, and collect evidence to test their ideas
We use a microscope to view very small things, such as cells	

Some living things are so small that we need a microscope to see them. Most of them are made of only one cell. These are microorganisms (or **microbes**). They are 'organisms' because they are living things, and carry out the seven life processes. They are 'micro' because they are very small.

Key term

microbe: another word for 'microorganism'.

Activity 2.1: Investigating microorganisms

What types of microorganisms are there?

A1 There are different types of microorganisms. Use different books and the internet to find:

   a) the names of the different types of microorganisms

   b) an example of each type

   c) an example of how microorganisms are useful

   d) an example of how microorganisms cause problems.

A2 Present your research as a table. Include in your table the place where you found each piece of information. You must include more than one place.

1 List *seven* processes that all microorganisms do.

2 What piece of apparatus (equipment) do you need to study microorganisms?

3 Give another word for 'microorganism'.

Fungi

Figure 2.1 shows a common type of **fungi**. These organisms may look like plants but they are not plants because they do not make their own food.

The type of fungi in figure 2.1 are large and contain many cells. Some fungi are much smaller, such as those found on mouldy bread. The **mould** is a fungus.

2.1 *Fungi, found in South Africa. People cook with some fungi but others are poisonous!*

2.2 *When a mould starts to grow, you cannot see it. As it grows, the number of cells it contains increases, which means you can see it.*

Mould fungi reproduce using **spores**. These are single cells made by the fungus. Spores are very light and so travel easily in the air. When they land on a source of food, they grow into new moulds. Fungi grow best in damp and warm places.

Some fungi are even smaller than moulds and have only one cell. These fungi are called **yeasts**.

4 a) What type of organism is a mould?

b) What is the mould in figure 2.2 feeding on?

c) What is the name of the life process that involves feeding?

d) Describe how this mould started to grow on this food.

e) State the conditions in which the mould will grow fastest.

5 Give *one* difference between fungi and plants.

6 a) Give *one* similarity between yeasts and moulds.

b) Give *one* difference between yeasts and moulds.

Key terms

fungus: type of organism that is not a plant or an animal. The plural is 'fungi'.

mould: fungus that decays things.

Key terms

spore: single cell released into the air by a fungus and which is able to grow into a new fungus.

yeast: type of fungus with only one cell.

2.3 *Yeast cells, magnified roughly 3000 times. We often write '3000 times' as × 3000.*

Activity 2.2: Investigating mould

What does mould look like?

A1 Write a plan for how to study mould cells. Include:

  a) a list of the apparatus (equipment) you need

  b) instructions on how to use the apparatus.

A2 If you study mould, make a drawing of its cells. A fungus cell contains cytoplasm, a nucleus and a cell membrane. It also has a cell wall. But unlike plant cells it never has a large vacuole or chloroplasts. You may not be able to see the nuclei.

Bacteria

The large image shown at the start of Chapter 2 shows some bacteria. A different type of **bacterium** is shown in figure 2.4. Bacteria have just one cell, although when they reproduce the new cells may stick to each other for a while. Bacteria are another different type of organism (they are not animals, plants or fungi).

Key term

bacterium: type of one-celled organism that is not a plant or animal or fungus. The plural is 'bacteria'.

7 **a)** Reproduction is a life process. What happens in reproduction?

  b) How do you know that the bacteria in figure 2.4 have reproduced?

8 The cells in figure 2.4 are magnified × 4000. What does this mean?

9 Which do you think are smaller, bacteria or yeasts? Give your reasoning.

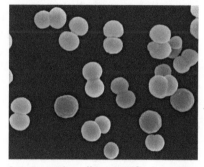

2.4 Bacteria cells, magnified × 4000.

Viruses

Viruses are tiny particles that get into the living cells of other organisms. A virus particle causes a cell to make copies of the virus. On their own, viruses are not alive and cannot reproduce.

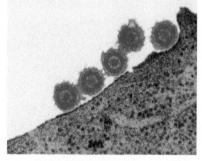

2.5 Virus particles, magnified × 37 000.

10 Which do you think are smaller, bacteria or viruses? Give your reasoning.

11 Why are viruses not like all other organisms?

Key term

virus: particle that is only alive when inside a living cell and cannot reproduce.

12 **a)** Use a ruler to measure the widest point across one of the yeast cells in figure 2.3. Use the magnification to work out the cell's size in real life.

b) Work out the size of a bacterium in figure 2.4.

c) Work out the size of a virus in figure 2.5.

13 **a)** The images on these pages are not taken with a light microscope. Find out what kind of microscope has been used.

b) Why has this microscope been used and not a light microscope?

Key facts:

✔ Microorganisms are living things that you need a microscope to see.

✔ Most microorganisms have only one cell.

✔ The three main types of microorganism are viruses, bacteria and some fungi (called yeasts).

Check your skills progress:

I can find and use information from different sources.

I can choose equipment to use and use it correctly.

I can make careful observations and measurements.

Louis Pasteur

Learning outcomes

- To describe how microorganisms spoil food, including the work of Louis Pasteur
- To describe stages of a scientific investigation, including:
  - asking scientific questions
  - making and explaining predictions
  - using evidence to make conclusions

Starting point

You should know that...	You should be able to...
Microorganisms carry out all seven life processes	Explain how scientists think up ideas, make predictions, and collect evidence to test their ideas
Microorganisms include bacteria, which have only one cell	

Microorganisms were first seen by the Dutch scientist Antonie van Leeuwenhoek in 1674 (see Chapter 1, Topic 5). For a long time, many scientists thought that microorganisms were created by the substances they were found in. For example, they thought that bread created mould. This idea was called 'spontaneous generation'.

Louis Pasteur was a French scientist who lived in the nineteenth century. He wondered why fresh, clear soup always went cloudy and started to smell bad. He used a microscope to look at soups. He found that cloudy, bad soup contained many microorganisms, which were not in fresh soup. He had an idea that microorganisms from the air landed in the soup. As they reproduced, they made the soup 'go bad'.

> **Key term**
>
> **Louis Pasteur**: French scientist who discovered that microorganisms spoil food.

1. Use the evidence in figure 2.6 to say which of the soups has 'gone bad'.

2. Describe *two* nineteenth century ideas to explain why soup goes bad.

Scientific method

The **scientific method** is the series of stages that scientists use in their investigations. They ask questions and think of

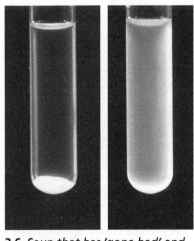

2.6 *Soup that has 'gone bad' and fresh soup.*

ideas to answer them. Then they plan experiments and make **predictions** about what will happen. They use the results from their experiments as **evidence** to make **conclusions**.

A **scientific question** is answered using an experiment. Pasteur asked: Do microorganisms from the air cause clear soup to go bad?

He planned an experiment using clear soup in glass containers. Some containers were open at the top. The tops of the other containers had an S-shaped tube. He boiled the soups in the containers and then left them.

A prediction says what you think will happen. Pasteur predicted that only the soup in the open-topped containers would go cloudy.

Scientists *explain* their predictions. Pasteur thought that microorganisms would fall into the containers with open tops and make the soup go bad. But the S-shaped tubes would trap microorganisms. This would stop the microorganisms reaching the soup in those containers, and so the soup would not go bad.

After a few weeks, only the soup in the open-topped containers was cloudy. Figure 2.8 shows one of Pasteur's S-shaped containers. Its soup is still clear!

1. Scientists ask a scientific question.

2. They plan an experiment.

3. They make a prediction.

4. They explain the prediction.

5. They collect results and consider them.

6. Finally, scientists use the evidence to make a conclusion.

2.7 *Pasteur's experiment shows the stages of a scientific investigation.*

Evidence is information that helps you decide if an idea is correct. Pasteur made the conclusion that his evidence showed he was correct – microorganisms from the air cause foods to spoil ('go bad').

3 Explain why 'Shall we paint the classroom blue?' is not a scientific question.

4 Describe Pasteur's evidence.

5 For each sentence, say whether it is a scientific question, a prediction, evidence or a conclusion.

a) If I add more fertiliser to plants, then they will grow taller.

b) The plants given the fertiliser were taller than the others.

c) Do plants grow taller with fertiliser?

d) Fertiliser makes plants grow taller.

Activity 2.3: Investigating Joseph Lister

How did Joseph Lister use Pasteur's ideas to develop modern surgery?

Joseph Lister was a Scottish surgeon who found a way of stopping infections after operations.

A1 Use books and the internet to:

a) suggest the scientific question that Lister asked

b) discover how he tested his idea

c) suggest what prediction he made

d) discover the evidence he used to make a conclusion.

A2 Present your research as a short report. Do not use more than 200 words. Record your sources of information.

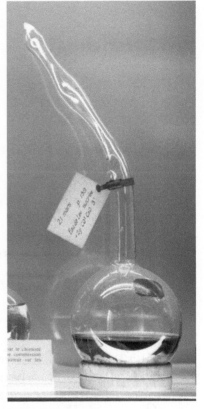

2.8 *One of Pasteur's original containers.*

Preserving foods

Microorganisms need water to survive. For thousands of years people have dried food to stop it spoiling. Pasteur discovered why this worked.

Other traditional ways of preserving food include adding lots of salt or sugar, and soaking foods in vinegar. All these methods kill the microorganisms.

Today, we use fridges and freezers to keep food cold. The cold stops microorganisms growing quickly.

To preserve some things, such as milk and other drinks, people heat them to a high temperature for a short time and then cool them quickly. The high temperature kills most of the microorganisms. Louis Pasteur invented this process, which we call pasteurisation.

Ghee

Ghee is a traditional ingredient in Arabic and South Asian cooking. It is made by boiling butter and collecting the clear liquid. This process kills microorganisms and removes the substances that they feed on. So, ghee lasts a very long time.

2.9 *Ghee.*

6 Explain why jam does not 'go bad' very quickly.

7 In some African countries, such as Nigeria, cassava is made into a mash and dried to form 'garri'. Explain why 'garri' keeps for a long time.

8 Explain why milk in a fridge lasts for a long time but still 'goes bad' eventually.

Key facts:

✔ Louis Pasteur did experiments to gather evidence to support his idea that microorganisms spoil food.

Check your skills progress:

I can identify scientific questions and predictions.

I can identify the evidence used to make conclusions.

I can recall the different stages of a scientific investigation.

Infectious diseases

Learning outcomes
- To describe what an infectious disease is
- To explain the causes of some infectious diseases
- To explain how we prevent some infectious diseases

Starting point

You should know that...	You should be able to...
Foods spoil because of the activities of microorganisms	Recall the different stages of a scientific investigation
	Identify scientific questions, predictions, evidence and conclusions

In the nineteenth century, many scientists thought that breathing in 'bad air' caused diseases. Louis Pasteur showed that microorganisms cause many diseases.

Symptoms are the effects of a disease on your body. A doctor uses symptoms as evidence, to make a conclusion about what disease someone has. This conclusion is a **diagnosis**. Table 2.1 shows the symptoms of some diseases and the microorganisms that cause them.

Key terms

diagnosis: saying what disease someone has.

fever: high body temperature.

symptom: effect of a disease on the body.

Disease	Caused by...	Symptoms
athlete's foot	fungus	red and itchy skin between the toes
chicken pox	virus	**fever**, raised red spots with yellow tops
cholera	bacterium	vomiting, very bad diarrhoea, muscle cramps
colds and influenza (flu)	virus	fever, sore throat, aches
food poisoning	bacterium	vomiting, stomach pain, diarrhoea
impetigo	bacterium	blisters on the skin that leave a yellow crust
measles	virus	fever, flat red spots
tuberculosis (TB)	bacterium	fever, coughing up blood

Table 2.1 *Some common diseases, their causes and symptoms.*

Using symptoms to diagnose disease

Three thousand years ago, Esagil-kin-apli wrote one of the first books explaining how to use symptoms to diagnose diseases. Esagil-kin-apli was the chief scholar to the King of Babylon, in modern-day Iraq.

1 **a)** Name a disease caused by a virus.

b) Name a disease caused by a bacterium.

2 Which disease has a symptom of coughing up blood?

3 A person with one of the diseases in Table 2.1 has red spots. What other information would a doctor need, to diagnose the disease?

4 Look at figure 2.10 and give a diagnosis.

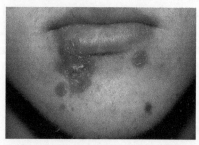

2.10 *The symptoms of many diseases include blisters, spots or a rash.*

Spreading diseases

Infectious diseases are diseases that spread from person to person. Microorganisms move from person to person in many ways. When the microorganisms start to grow and reproduce in a person, that person is **infected**.

5 Why are diseases caused by microorganisms 'infectious'?

The air can carry microorganisms from person to person. When people cough or sneeze, they spray tiny droplets of liquid into the air. These droplets contain microorganisms. If someone breathes in the droplets, they could become infected. Colds, measles, chicken pox and tuberculosis spread in this way.

Touching spots or blisters on the skin can spread microorganisms, such as those that cause impetigo and chicken pox. Athlete's foot spreads when people touch things that have been in contact with an infected foot (such as a wet floor near a swimming pool).

Some microorganisms spread in foods and drinks. For example, cholera bacteria spread in water.

Animals spread some diseases. A bite from a dog with rabies transfers the rabies virus. Mosquitoes spread the microorganism that causes malaria.

6 **a)** State the name of an infectious disease.

b) Why is this disease 'infectious'?

7 State an example of a disease spread by animals.

8 Copy the diseases from Table 2.1. Write down how each disease spreads.

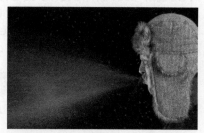

2.11 *Special photography shows the droplets of liquid in a sneeze. In some sneezes, the droplets travel at 100 km/h!*

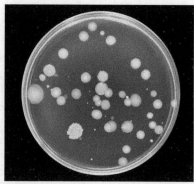

2.12 *This dish contains nutrient agar, which bacteria feed on. As they grow and reproduce, the bacteria form clumps (called colonies). This dish contains bacteria from a small drop of sea water.*

Stopping diseases from spreading

To stop people from getting infectious diseases we prevent the spread of microorganisms. Using a tissue when sneezing and covering your mouth when coughing are simple ways to stop microorganisms from spreading through the air.

We stop food and drink from causing diseases in many ways. Preserving food by adding a lot of salt, sugar or vinegar kills bacteria and fungi. Cooking foods properly destroys the microorganisms in them. Pasteurisation kills bacteria and fungi in drinks, such as milk.

Many of the products we put on our skins contain **antiseptics** to kill microorganisms.

Disinfectants kill microorganisms on surfaces that we touch. We also move sewage away from people, and treat it to kill microorganisms. Cholera often spreads when sewage gets into drinking water.

Injecting people with a **vaccine** prevents some diseases.

9 Why does cooking stop people from getting food poisoning?

10 Explain why you should use a tissue to sneeze into if you have a cold.

11 List some ways that have stopped microorganisms from infecting you today.

Key terms

antiseptic: substance that kills microorganisms but is safe for us to put on our skins.

disinfectant: substance that kills microorganisms on surfaces that we touch.

vaccine: substance injected into people to stop them getting an infectious disease.

Preventing cholera

John Snow was an English doctor. In 1854 he wanted to know why people in London were dying of cholera. He recorded where people with cholera lived and noticed that many of them had homes near a certain water pump (where they collected drinking water). This gave him an idea. He thought that if people stopped using this pump, then the number of people with cholera would decrease. The pump handle was removed, so people had to get water from other pumps. After this, the number of people with cholera decreased rapidly.

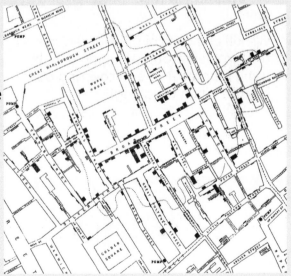

2.13 *John Snow recorded his observations on a map. Each spot shows a death from cholera.*

How did Robert Koch use Pasteur's ideas to find out more about infectious diseases?

Robert Koch was a German scientist, who read Louis Pasteur's work and then started to investigate microorganisms.

A1 Use books and the internet to:

    a) discover *one* method that Koch invented to study bacteria

    b) find the names of *three* bacterial diseases that Koch investigated.

A2 Record your sources of information.

12 Read the passage about John Snow on page 47.

    a) Suggest a scientific question he might have asked.

    b) What was his prediction?

    c) Suggest the conclusion that John Snow made.

    d) What evidence did he use to make his conclusion?

13 **a)** Describe how cholera is spread from one person to another.

    b) Suggest *two* ways in which we stop this disease spreading.

Key facts:

✔ Some microorganisms cause infectious diseases.

✔ Microorganisms spread through the air, by touch, in food, in water, and by animals.

✔ We kill and control microorganisms to stop them from spreading (e.g. by using disinfectants and covering our mouths when coughing).

Check your skills progress:

I can ask scientific questions.

I can make predictions.

I can use evidence to make conclusions.

Useful microorganisms

Learning outcomes
- To describe some uses of microorganisms

Starting point

You should know that...	You should be able to...
Microorganisms carry out all seven life processes, including respiration	Describe how scientists plan investigations to answer their questions
Foods spoil because of the activities of microorganisms	

We use bacteria and fungi to make many foods. Figure 2.14 shows some examples.

1. List the foods you often eat that are made using microorganisms.

2. Give the name of a microorganism that is not used to make food.

2.14 *Cheese and bread are two of the many foods made using microorganisms.*

Making bread

We use yeast to make some types of bread. Yeast cells feed on sugar in bread dough. As they respire, the yeast make bubbles of carbon dioxide gas. This gas causes the dough to rise up and increase in height. Then we bake the dough to make bread.

Key term

variable: something that may change.

Activity 2.5: Investigating yeast

In what conditions do yeast grow best?

A bread dough contains 100 g of flour, 10 g of sugar, 120 cm³ of water and 7 g of yeast. Some people add more sugar and some add less. Not everyone leaves the dough at the same temperature or for the same time.

A1 A **variable** is something that may change. List the variables for bread making.

A2 Write a scientific question using *one* variable. Start it as 'What happens if we change...?'

A3 In an experiment, you change only *one* variable. You keep the other variables the same. Which variables would you keep the same?

A4 Which variable would you measure?

A5 Predict what will happen, and explain your prediction.

A6 How would you use evidence from this experiment to make a conclusion?

3 What type of microorganism is yeast?

4 A student made six batches of bread dough, and added different amounts of sugar to each. The student left each dough to rise for 20 minutes and then measured the increase in volume. The table shows the results.

Mass of sugar added (g)	Increase in volume after 20 minutes (cm³)
0	11
2	21
4	25
6	29
8	35
10	37

a) Which variable did the student change?

b) State *two* variables that should be kept the same.

c) Which variable did the student measure?

d) Use the evidence in the table to make a conclusion.

Decay

During **decay**, materials break into smaller parts. **Decomposers** are microorganisms that cause decay. Bacteria and fungi are decomposers. Decay is not useful for food but is important for removing dead organisms and animal wastes. Decay releases useful substances from dead organisms and wastes, for other organisms to use.

5 a) What is a decomposer?

b) Explain why decomposers are important.

6 A scientist took four containers, each containing 100 g of dead plants. The scientist added decomposers to each container and put them at different temperatures for 30 days. The table shows the results.

Temperature (°C)	Mass of plant material after 30 days (g)
5	97
15	94
20	82
30	70

2.15 *Decomposers on animal droppings.*

Key terms

decay: when materials break into smaller parts. Microorganisms often cause this.

decomposer: microorganism that causes decay.

a) Which variable did the scientist change?

b) State *two* variables that should be kept the same.

c) Which variable did the scientist measure?

d) Use the evidence in the table to make a conclusion.

7 Unleavened bread, such as Al Jabab bread eaten in the United Arab Emirates, is made without yeast. Suggest a difference between this bread and the bread in figure 2.14. Explain your reasoning.

Key facts:

✔ Bacteria and fungi are useful for making food (e.g. yeast is used to make some types of bread).

✔ Bacteria and fungi are decomposers. They decay and remove dead organisms and animal wastes. This releases useful substances, which other organisms are then able to use.

Check your skills progress:

I can plan experiments.

I can identify variables, including those that you change, those that you keep the same and those that you measure in an experiment.

Quick questions

1. Milk can be pasteurised by heating it to 70 °C for 15 seconds. This is to:

    a make milk taste better   b turn the milk into cream

    c kill microorganisms   d turn the milk sour   [1]

2. The microorganisms used to make bread rise are:

    a yeasts   b bacteria

    c viruses   d eggs   [1]

3. An infectious disease is one that:

    a spreads through food   b spreads from person to person

    c cannot be treated   d causes a rash   [1]

4. Colds and influenza (flu) are caused by:

    a fungi   b yeasts

    c bacteria   d viruses   [1]

5. Which of these is a discovery made by Louis Pasteur?

    a Foods make microorganisms.

    b Microorganisms from the air cause foods to go bad.

    c Microorganisms only reproduce when inside other living cells.

    d Microorganisms are not living things.   [1]

6. Explain why we wash our hands with soap after using the toilet.   [2]

7. Louis Pasteur asked: 'Do microorganisms from the air cause clear soup to go bad?' Why is this a scientific question?   [1]

8. (a) Describe *one* use of bacteria.   [1]

    (b) Describe *one* way in which bacteria are not useful.   [1]

9. (a) Explain why dried foods last a very long time.   [2]

    (b) Explain why jam lasts a very long time.   [2]

10. For each sentence, say whether it is a scientific question, a prediction, evidence or a conclusion.

   (a) Do yeast cells reproduce faster when given more sugar? [1]

   (b) If I add more sugar to the water, then the yeast will reproduce faster. [1]

   (c) There were more yeast cells found in the water with more sugar. [1]

   (d) Increasing the amount of sugar makes yeast cells reproduce faster. [1]

11. (a) What is a decomposer? [1]

   (b) Name *two* types of decomposers. [2]

   (c) As decomposers feed on wastes, they respire. What gas do they make? [1]

   (d) Describe *one* way in which decomposers are useful. [1]

   (e) Describe *one* way in which decomposers are not useful. [1]

Connect your understanding

12. Manure is rotting animal waste mixed with parts of dead plants. Explain why manure helps crops to grow. [2]

13. (a) Figure 2.16 shows a yeast cell. What are parts X, Y and Z? [3]

   (b) Give *one* way the yeast cell is similar to a plant cell but not to an animal cell. [1]

   (c) What green structures are in some plant cells but not in yeast cells? [1]

   (d) What group of organisms do yeast belong to? [1]

   (e) The image is magnified × 400. What does this mean? [1]

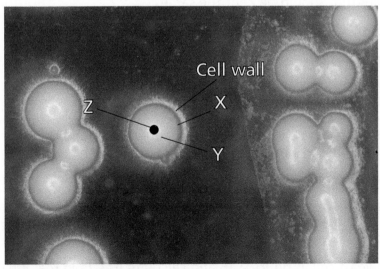

2.16 *Yeast cells (magnified × 400).*

14. If floods hit an area, sewage can get into the water supply.

    (a) Why will boiling water before drinking it prevent diseases? [1]

    (b) Give the name of *one* disease that is spread by dirty water. [1]

15. A carton of milk is on top of a fridge. Another carton of milk is inside the fridge.

    (a) Predict which milk will spoil first. [1]

    (b) Explain your prediction. [1]

16. Mumps is a disease caused by a virus. It causes parts of your mouth to swell up and it is painful to swallow. Some people receive an injection to stop them getting this disease. What does the injection contain? [1]

17. A student made six batches of bread dough, and left each at a different temperature for 30 minutes. The student then measured the increase in height of each dough. The table shows the results.

Temperature (°C)	Increase in height after 30 minutes (mm)
5	0
10	2
15	9
20	19
25	28
30	35

    (a) Explain why the height of the dough increased. [1]

    (b) Which variable did the student change? [1]

    (c) State *two* variables that should be kept the same. [2]

    (d) Which variable did the student measure? [1]

    (e) Use the evidence in the table to make a conclusion. [2]

Challenge questions

18. A baker added some yeast cells to water containing sugar. The number of yeast cells increased for 10 hours, and then stopped increasing.

    (a) Which life process caused an increase in the number of cells? [1]

    (b) Suggest why the number of cells stopped increasing. [1]

19. Some scientists do not think that viruses are organisms. Explain why not. [2]

20. Measure the labelled yeast cell in figure 2.16 and calculate its size in real life. [2]

3

Chapter 3
Habitats and the environment

What's it all about?

Jerboas are small animals that live in deserts, in Africa and Asia. Their fur is the colour of sand. This means that they blend with their surroundings and so predators are less likely to see them. Jerboas are active at night and so have large eyes to see well in moonlight. Their large, hairy feet stop them sinking into the sand. They can also survive without drinking water (they get their water from their food). These 'adaptations' allow jerboas to survive in deserts.

You will learn about:
- How organisms survive in the places where they live
- How organisms depend on each other
- How humans affect the environment

You will build your skills in:
- Using information to make conclusions
- Recording observations accurately in different ways
- Using evidence to support ideas

Adaptations of organisms

Learning outcomes
- To describe the places where organisms live
- To use sampling to research the types of organisms in an area
- To describe how organisms are adapted to where they live

Starting point

You should know that...	You should be able to...
Organisms carry out seven life processes	Present data using tables

Habitats

The place where an organism lives is its **habitat**. Some habitats are large, such as a desert. Some habitats are small, such as a pond.

The things that organisms need are **resources**. Habitats provide organisms with resources, such as:

- water
- shelter and protection
- food.

There are many types of organisms in a habitat. A habitat also contains non-living parts, such as temperature, light, wind, water and rocks. These are **physical factors**.

All the organisms and all the physical factors in a habitat form an **ecosystem**.

The surroundings of an organism are its **environment**. An environment contains:

- other organisms
- a range of physical factors.

1 a) What is a habitat?

  b) Name *two* habitats.

  c) Name *one* habitat in which fish live.

2 a) Name the habitat in figure 3.1.

  b) List the different organisms living there.

Key terms

ecosystem: all the organisms and the physical factors in a habitat.

environment: the other organisms and physical factors around an organism.

habitat: the place where an organism lives.

physical factor: non-living part of an environment (e.g. wind).

resource: anything that is needed or used by an organism.

3.1 *This savanna habitat is in Kenya. A savanna is an area of open grassland with some trees.*

c) Look at the giraffes in figure 3.1. Describe their environment.

3 Describe your environment now.

4 Describe *two* physical factors in a rainforest habitat.

5 Fungi live on dead organisms and waste. Most fungi live in dark, warm and wet conditions. Suggest a habitat in which fungi live.

Sampling organisms in a habitat

To discover what lives in a habitat, scientists examine small parts of it. These small parts are **samples**. The photos show some ways of collecting samples.

3.2 *A **pitfall trap** is a container buried in the ground, with a cover to stop rain getting in. It traps animals that fall into it. Pitfall traps have different sizes.*

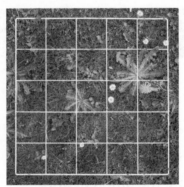

3.3 *A **quadrat** is a square frame. Scientists place it in an area and look for the different organisms inside it.*

3.4 *Scientists collect samples from water using jars or nets. Scientists also sweep some nets through tall grasses to collect small animals living there.*

To examine small animals, scientists need to handle them carefully and look at them closely. The photos below show some ways to do this.

3.5 *If you carefully suck on one tube of a **pooter**, small animals are sucked into the collecting jar.*

3.6 *To examine and count small, crawling animals without harming them, we use soft brushes to move them.*

3.7 *We magnify small animals and parts of plants with a **magnifying glass** (or **hand lens**).*

6 Why do scientists use soft brushes to handle small organisms?

7 What type of organism does a pitfall trap collect?

8 Why do we use a magnifying glass to examine small animals?

9 Describe a type of animal that you could find out about using a quadrat.

Presenting results

Scientists took samples of a savanna in East Africa to discover which grasses lived there. In the results table below, the ticks show the grasses found in each sample.

Type of grass	Sample number											
	1	2	3	4	5	6	7	8	9	10	11	12
elephant grass	✓	✓	✓			✓		✓		✓		
pan dropseed												
red dropseed		✓		✓	✓							
red grass		✓		✓			✓		✓	✓	✓	✓

A bar chart makes it easier to compare the different grasses.

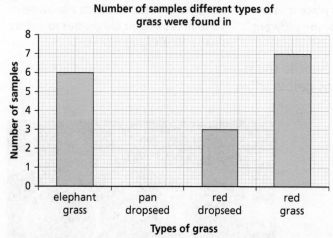

Number of samples different types of grass were found in

3.8 *Bar chart showing the number of samples that different grasses were found in.*

10 Look at the table and the bar chart.

  a) Which was the most common grass in this area?

  b) Which grass was not found in this area?

c) Suggest which method was used to collect these samples.

d) Why is it better to show the results in a bar chart rather than in a table?

e) A line graph is *not* a good choice to show these results. Why is this? (Hint: Look back at Topic 1.1.)

11 In part of the Amazon rainforest in Brazil, some students left pitfall traps for one night. The table shows the number of beetles in each trap the next morning. Present these results using a bar chart.

Type of beetle	Pitfall trap number									
	1	2	3	4	5	6	7	8	9	10
C Coprophanaeus				✓					✓	
D Dichotomius	✓		✓	✓	✓	✓		✓	✓	✓
E Eurysternus										
O Onthophagus		✓		✓			✓		✓	✓

12 Look at figure 3.1 on page 56.

a) Draw a table to show the number of each different animal in the photo. Only include those animals that you can clearly see.

b) Present your results as a bar chart.

c) A scientist says that the giraffe is the most common animal in this habitat. Does the evidence in your table and bar chart support this conclusion?

What types of organisms live in a habitat near you?

A1 Choose a habitat near you and describe it. You could include a drawing.

A2 Make a prediction about what organisms live there.

A3 Decide how you will take samples from your habitat.

A4 Carry out your investigation. You may need to use books (field guides) to help you identify some of the organisms that you find.

A5 Present your results as a table (and as a bar chart, if you can).

A6 Make a conclusion about which organisms are the most and least common.

Adaptations

Organisms have features that allow them to survive in their habitats. We call these features adaptations and say that organisms are adapted to where they live.

Fish are adapted to water by having fins to swim. They also have gills, which they use to take oxygen from the water. Fish cannot breathe out of the water. They are *not* adapted to living on land.

3.9 *This Mandarin duck is from China. Its beak is an adaptation that allows it to catch insects.*

3.10 *Small plants in rainforests are adapted to dark, shady areas.*

13 **a)** How are ducks adapted to swimming on water?

b) Suggest how ducks are adapted to surviving attacks by other water animals, such as crocodiles.

14 Suggest how a lion is adapted to eating meat.

15 Look at figure 3.10. Explain how small plants are adapted to this habitat.

16 **a)** Give the name of *one* animal in your country.

b) Explain *one* of its adaptations.

Some organisms have adaptations that let them live in habitats where the physical factors make it difficult to live. These habitats include deserts (where there is little water) and the Arctic (where it is very cold).

Thick fur helps stop heat escaping (to help keep the bear warm)

Small ears help stop heat escaping

Rough soles on its feet grip the ice

White fur helps stop the bear being seen by the animals that it hunts

Large feet help stop the bear sinking into the snow and are good for swimming

3.11 *Polar bears are adapted to living in the Arctic.*

Spines

No leaves and so it loses very little water in the hot sunshine (plants lose water through their leaves)

Water stored in the stem

Roots cover a large area to absorb water very quickly when it rains

3.12 *Cacti are adapted to living in deserts.*

17 Explain how polar bears survive the cold of the Arctic.

18 Suggest why the spines on a cactus plant help it to survive.

19 Oryx are antelopes that only sweat when the temperature goes above 46 °C. Explain how this adaptation allows oryx to survive in a hot, dry desert.

20 Design an animal that is adapted to living in a very cold, dry sandy desert.

Are leaves in shady places larger than leaves in sunny places?

A student thinks that leaves that are always in the shade on a plant will be bigger than those that are usually in the sunshine. Plants need light to make food in their leaves, and bigger leaves trap more light. The student makes this prediction:

If the amount of light is less, then the leaves will be bigger.

You are going to choose a tree or plant in your local area and test this prediction.

A1 Plan a way to look at the leaves in the different places.

A2 How will you stay safe when you look at the leaves? For example, it may be dangerous to climb trees.

A3 How will you compare the sizes of the leaves?

A4 A variable is something that may change. In an experiment, you change only one variable. You measure another variable. You keep all the other variables the same.

   a) Which variable are you changing?

   b) Which variable are you measuring?

   c) Suggest *one* variable that you are keeping the same.

A5 Carry out your investigation.

A6 Present your results as a table.

A7 Make a conclusion, saying whether the prediction is right or wrong.

Adaptations for daily changes

Physical factors in a habitat change during a day. These are **daily changes**. Organisms have adaptations to survive daily changes.

Some flowers close at night to keep their pollen safe. They open again during the day, when the insects that collect their pollen are active.

Nocturnal animals are active at night. They often have very good eyesight, to see things in moonlight. They may also have excellent hearing. Some nocturnal animals are **predators** and use their hearing to hunt their **prey** (animals that they eat). The prey animals use their hearing to avoid their predators.

Animals living on the seashore are adapted to the tides. Sea anemones use soft, fine tentacles to feed under water. When the tide is out, they pull in their tentacles so that they do not dry out.

Key terms

daily change: change in physical factors during the course of a day.

nocturnal: active at night.

predator: animal that hunts and eats other animals (called prey).

prey: animal that is hunted and eaten by other animals (called predators).

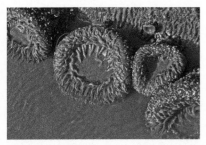 

3.13 *Sea anemones are adapted to the tides.*

21 Most owls are nocturnal.

    a) What does this mean?

    b) Explain why owls have very large eyes.

22 Look at the table in question 11 on page 59. Suggest which animal is *not* nocturnal. Explain your reasoning.

23 Explain why nocturnal mice have excellent hearing.

24 Porcupines are nocturnal animals that have a covering of spines. Suggest why they have spines.

25 How does a sea anemone protect itself when the tide goes out?

Adaptations for seasonal changes

Physical factors in a habitat change during a year. These are **seasonal changes**. Organisms have adaptations to survive the different **seasons**.

When growing conditions are good, many plants make seeds. The plants then die when conditions become harsh (for example, when it is very dry or cold). Their seeds survive in the ground ready to grow into new plants when conditions improve again.

Deciduous plants lose their leaves during certain seasons. Without leaves they do not need much water. It is difficult for plants to get water during dry seasons and cold seasons. In dry seasons there is not much rain and in cold seasons the water may be frozen in the ground.

Key terms

deciduous: plant that loses its leaves during a certain season of the year.

season: time during the year with a certain set of physical factors.

seasonal change: change in physical factors during the course of a year.

Many animals move when the seasons change. This is **migration**. Many butterflies fly to new areas just before the rainy season. They return when the rains have gone.

Many birds migrate to warmer areas when cold seasons arrive. This helps them to survive because the warmer areas contain more food (e.g. seeds, insects). If animals cannot find food they may starve and die.

Key term

migration: when animals move from one area to another as the seasons change.

Tracking bird migration

Amur falcons leave Mongolia and China when it starts to get cold in August and September. Large flocks of the birds migrate to India and then across the Arabian Sea to Southern Africa. In May, they make the 11 000 km journey back again.

Scientists working for organisations in Hungary, India and Abu Dhabi work together to track the birds. They catch some of the birds and attach small tracking devices.

26 a) How far do Amur falcons fly each year?

b) Explain why the falcons leave China in August.

c) What word describes this type of journey?

27 Suggest why butterflies leave an area before heavy rains arrive.

28 Flesh-footed shearwater birds leave Australia in May. Many fly to Oman. Suggest a change in a physical factor that causes them to leave Australia.

Other animals have adaptations that allow them to stay in their habitats.

Stoats are small animals that hunt mice. They live in many northern parts of the world. They have brown fur, to blend in with the soil and plants. They have white fur in winter, to blend in with the snow. This means that mice are less likely to see them.

In the Himalayas, Asian black bears eat a lot before the cold, winter season. They store fat in their bodies, to use for respiration. Then they find a safe place to hibernate until warmer temperatures return. During **hibernation**, an animal does not move much, and its body temperature, breathing rate and heart rate all decrease. Hibernating animals do not eat during cold months when there is little food.

Key term

hibernation: when animals go into a type of sleep during cold seasons.

29 African bullfrogs live in savanna habitats. In the hot, dry season they bury themselves, and wait until the rain comes again. What might happen to African bullfrogs if they did not bury themselves?

30 Explain why Asian black bears eat a lot before the cold season.

31 a) Dormice hibernate. Describe what the dormice do.

b) How does this help them to survive?

32 a) In countries that have cold winters, the water in the ground freezes. Suggest why many trees in these countries are deciduous.

b) Teak trees grow in India, Myanmar and Thailand, in places where it does not get cold. Explain why teak trees are deciduous.

c) Teak trees have flowers from June until August. Suggest why they flower at this time of year.

Key facts:

✔ Organisms live in many different habitats.

✔ An organism's environment is the physical factors and other organisms around it.

✔ An ecosystem is all the organisms and physical factors in a habitat.

✔ Adaptations help organisms survive in their environments and habitats.

✔ Adaptations help organisms survive changes in their habitats (such as daily and seasonal changes).

Check your skills progress:

I can choose the correct equipment to take samples from a habitat.

I can present information using tables and bar charts.

I can use evidence to support my conclusions.

Food chains

Learning outcomes
- To recall the words we use to describe how organisms feed
- To explain how organisms depend on one another for food
- To draw food chains

Starting point

You should know that...	You should be able to...
Adaptations help organisms survive in their habitats, including avoiding being eaten or helping to catch prey	Show a sequence using a flow chart

Organisms need food for energy. They release the energy using respiration.

Plants produce their own food. They are **producers**. Animals need to consume (eat) other organisms. They are **consumers**.

Animals that only eat plants are **herbivores**. Those that only eat other animals are **carnivores**. Those that eat plants and animals are **omnivores**.

1 Which life process do organisms use to release energy?

2 Is a tree a producer or a consumer? Give a reason for your choice.

3 a) Name a herbivore in your country.

b) Name a carnivore in your country.

c) Use *two* of the **bold** words above to describe yourself.

Key terms

carnivore: animal that eats other animals.

consumer: animal that eats other living things.

food chain: list with arrows that shows what eats what in a habitat.

herbivore: animal that eats plants.

omnivore: animal that eats both plants and animals.

producer: organism that makes its own food, such as a plant.

A **food chain** shows what eats what in a habitat. A food chain starts with a producer.

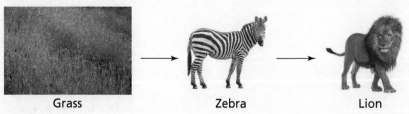

Grass   Zebra   Lion

3.14 *A food chain from an African savanna.*

4 Look at figure 3.14 and identify a:

a) producer

b) consumer

c) carnivore

d) herbivore

e) predator

f) prey.

5 In a rainforest in Borneo, grey tree rats feed on fig trees. Borneo pythons eat grey tree rats.

a) Write a food chain for these organisms. (Do not draw pictures.)

b) Underneath the names of the organisms, write one or more of these words:
carnivore   consumer   herbivore
predator   prey   producer

c) Draw a table to show the information in your food chain. Give your table column headings:
Organism   Type of nutrition   Where it gets
            (e.g. consumer)   its energy from

d) What is a Borneo python's prey?

6 a) Jerboas in the Gobi Desert eat plant leaves, seeds and insects. Which of these words best describes a jerboa?
carnivore   herbivore   omnivore

b) Why did you choose this word?

The first consumer in a food chain is the **primary consumer** and the second is the **secondary consumer**. The last consumer in a food chain is the **top predator**.

The arrows in a food chain show energy flow. Energy from the Sun (trapped by producers) flows to the primary consumer. Energy in the primary consumer flows to the secondary consumer.

7 Look at this food chain:

sagebrush → Tolai hare → steppe eagle

Identify a:

a) producer

b) secondary consumer

c) top predator.

Key terms

primary consumer: the first consumer in a food chain, which is always a herbivore.

secondary consumer: the second consumer in a food chain, which is always a carnivore.

top predator: the last predator in a food chain.

8 **a)** Where do plants get their energy from?

b) Where does a secondary consumer get its energy from?

9 The third consumer in a food chain is the tertiary consumer. The fourth one is the quaternary consumer.

In a grassland habitat, aphids feed on grass. Sparrowhawks hunt smaller birds, such as thrushes. Aphids are prey for spiders. Thrushes are predators of spiders.

a) Write a food chain for these organisms.

b) Label the organisms in your chain with words to describe their nutrition.

Key facts:

✔ A food chain shows what eats what in a habitat.

✔ There are scientific words to describe the position of each organism in a food chain and its nutrition (carnivore, herbivore, omnivore, predator, prey, producer, primary consumer, secondary consumer, top predator).

Check your skills progress:

I can use information to draw food chains.

Human impact on the environment

Learning outcomes
- To describe ways that humans damage habitats
- To describe ways that humans help habitats
- To identify advantages and disadvantages of different energy resources

Starting point

You should know that...	You should be able to...
Organisms are affected by changes in the physical factors in their habitats	Explain how scientists think up ideas, make predictions, and collect evidence to test their ideas
We burn a lot of fossil fuels (such as oil, coal and natural gas) to power vehicles and produce electricity	Use tables, bar charts and line graphs

Human population increase

The number of people living on Earth is the world human **population**. The line graph below shows how the world human population has changed over 2000 years.

Key term

population: the number of one type of organism in a place.

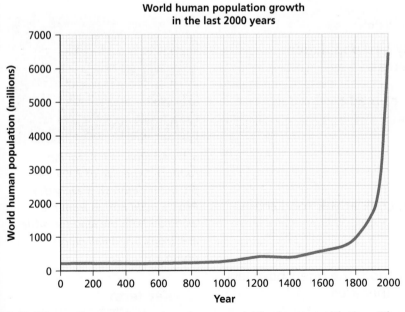

World human population growth in the last 2000 years

3.15 *We use line graphs to show how a variable changes with time. The variable in this graph is world human population.*

An increasing human population needs:

- more space to live in
- more food
- more water
- more resources from the Earth (such as oil and some rocks).

People clear natural habitats for buildings and farms. We also destroy habitats by building dams to form reservoirs, and by mining rocks used for buildings and roads.

When we destroy habitats, animals and plants may have nowhere to live, or they may not be able to get food. If we cut down the trees, many birds will have nowhere to build their nests. The animals that eat the trees will have no food and so may die.

1 Look at the line graph in figure 3.15.

   a) About how many people were in the world in 1800?

   b) What has happened to the human population since 1800?

   c) Why does an increase in the number of people cause more habitat destruction?

2 **a)** What is **deforestation**?

   b) Why do people deforest areas?

   c) Describe *one* reason why bird populations decrease in deforested areas.

3 Explain why the farmer has cleared the land in figure 3.17.

3.16 *We destroy habitats when towns and cities increase in size, and when we dig quarries. This is a limestone quarry.*

3.17 *An area of Amazonian rainforest the size of 2000 soccer pitches is deforested every day. People build homes from the wood and use the land for cattle and crops. The small plants in this photo are maize.*

Damaging food chains

Humans also damage food chains. Look at this food chain, from a forest habitat.

grass → spotted deer → tiger

If humans kill the spotted deer for food, the tigers have less food. They may die or move to another area.

If humans kill tigers to stop them killing cattle, the deer population increases. More spotted deer eat more grass, and there is less grass for other animals.

Key term

deforestation: cutting down forests.

4 Look at the food chain with the deer and the tiger.

a) Why will killing the deer reduce the numbers of tigers?

b) Why will killing tigers increase the deer population?

5 Look at this food chain from a lake.

algae → snail → lake herring → trout

Some people fish too many herring. How will this affect the numbers of snails and trout? Give reasons for each answer.

a) Snails.

b) Trout.

6 This food chain includes thrushes, which are omnivorous birds that eat insects and grains from wheat plants.

wheat plants → locust → thrush → sparrowhawk (leaves)

In the 1950s in China, the government told people to kill the thrushes to stop them eating stores of wheat grains. Explain why killing the thrushes meant that the people had even less wheat.

Pollutants

Pollution is when organisms are being harmed by a substance in their habitat. The substance causing the harm is a **pollutant**.

Human activities cause a lot of pollution. Pollutants include oil from tankers, poisonous substances from factories and substances we use to make our lives easier.

CFCs are substances we use in spray cans, fridges and air conditioners. In the 1970s, scientists found that CFCs were destroying the ozone layer in the atmosphere. This layer protects organisms from dangerous rays from the Sun. These rays may cause skin cancer. Countries around the world have now agreed to ban CFCs and so stop **ozone depletion**.

7 Would you use a bar chart or a line graph to show changes in CFC levels in the atmosphere over the years? Give a reason for your answer.

Key terms

ozone depletion: reducing the amounts of ozone.

pollutant: substance that causes harm to organisms.

pollution: when organisms are being harmed by a substance.

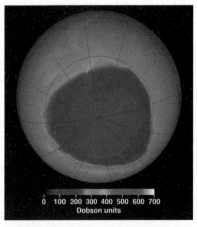

0 100 200 300 400 500 600 700
Dobson units

3.18 *This satellite image shows the thinnest parts of the ozone layer (over Antarctica) in blue and purple.*

8 Use figure 3.18 to suggest why more people get skin cancer in Australia than in countries in Europe.

9 The ozone layer was thinnest in 1994. Suggest why it is now thicker again.

Air pollution

Fuels are substances that release energy. We use them in factories, vehicles and power stations to make electricity. Most fuels formed over millions of years from dead plants and animals. These are **fossil fuels**. Oil, coal and natural gas are fossil fuels.

To release energy from fossil fuels, we burn them. This also makes carbon dioxide gas. In the atmosphere, this gas acts a bit like the glass on a greenhouse. It lets energy from the Sun reach the Earth and warm it up and then stops the energy escaping. Carbon dioxide helps to cause the **greenhouse effect** on Earth.

Scientists think that increasing amounts of carbon dioxide in the atmosphere make the Earth become warmer. This **global warming** may change the weather around the world (**climate change**).

10 Name *three* fossil fuels.

11 Name a gas that causes global warming.

12 Explain why some scientists think we should stop burning fossil fuels.

Many fossil fuels (such as diesel) release small smoke particles when they burn. These help to form **smog** (a chemical fog), which may damage people's lungs and make breathing difficult.

Activity 3.3: Investigating pollutant particles

Is there pollution caused by burning fossil fuels in your area?

Hang a white tissue or piece of fabric outside for a week. Then compare it with one that has been kept inside.

A1 Make a prediction and explain why you think this.

A2 Say whether your results match your prediction.

A3 Use your results as evidence to make a conclusion.

A4 What piece of equipment could you use to examine particles more closely?

A5 Present your findings as a short report. Do not use more than 200 words.

Key terms

climate change: changes to weather patterns.

fossil fuel: fuel such as coal, oil and natural gas, made over millions of years from dead organisms.

fuel: substance that releases energy.

global warming: increasing temperatures around the Earth and its atmosphere.

greenhouse effect: when gases in the atmosphere trap energy and cause the Earth to warm up.

smog: unpleasant chemical fog.

3.19 *Smog in Warsaw, Poland.*

Some fossil fuels (such as coal) contain substances that produce sulfur dioxide gas when burned. This gas dissolves in clouds to form an acid, which falls as **acid rain**. This may kill trees.

Vehicle engines that burn fuels make gases called 'oxides of nitrogen'. These gases also cause acid rain, and help to cause smog.

13 Draw a table to show some air pollutants and their problems.

14 **a)** Name the substances made by burning fossil fuels that form smog.

b) Why is smog a problem in cities?

15 Explain why using more electric cars may improve our cities.

3.20 *Acid rain killed these trees in Germany.*

Renewable energy resources

Fossil fuels will not last forever. We are using them up much faster than they can form, so they will run out. They are **non-renewable**. We need to use **renewable** energy resources (ones that will not run out).

Governments around the world have agreed to burn less fossil fuel. They are looking at alternative ways of making electricity and running vehicles.

We now use some plants and algae to make **biofuels** for vehicles. But this means farmers use land for growing biofuel plants instead of growing food for humans.

16 **a)** Why is coal a non-renewable fuel?

b) Why are biofuels renewable?

Many power stations use fossil fuels. Renewable ways of producing electricity are becoming more common. These do not release gases into the air and so cause less pollution. They include power from:

- wind
- waves
- tides
- the Sun (solar power)
- water flowing out of dams (hydroelectric power).

Key terms

acid rain: rain that is much more acidic than usual.

biofuel: fuel made using plants or algae.

non-renewable: something that will not last forever.

renewable: something that will not run out.

3.21 *Some aeroplanes now burn biofuels.*

17 Is wave power renewable or non-renewable? Give a reason for your answer.

18 **a)** Give *one* advantage of solar power compared to burning fossil fuels.

   b) Suggest *one* disadvantage of solar power.

19 **a)** Describe *one* way in which wind power is a benefit for the environment.

   b) Describe *one* way in which wind power may harm the environment.

20 Explain how using satellite tags to track migrating birds is useful for deciding where to build wind turbines.

3.22 *Wind turbines provide renewable electricity but they only produce electricity if there is wind. They may also be dangerous for migrating birds, they may damage habitats and some people think they are ugly.*

Activity 3.4: Investigating hydroelectric power

A country is deciding whether to build a hydroelectric power station. It involves building a dam across a river to create a reservoir.

The government wants you to 'evaluate' this idea. In an evaluation, you state at least one good point about an idea and at least one bad point. Then you decide if, overall, the idea is good or bad.

A1 Write a report about this idea. Set out your report like this:

- a title
- a sentence to explain what your report is about
- a new paragraph to explain one or more reasons against the dam
- a new paragraph to explain one or more reasons in favour of the dam
- a final paragraph to explain what you think and why.

A2 The government has asked you to make a presentation about your ideas. Your presentation should last no longer than 2 minutes. Your teacher may ask you to make a presentation to your class.

Helping the environment

Humans damage habitats but we help them too.

Scientists are developing ways of causing less pollution and removing pollutants from the environment. This helps more organisms to survive.

To protect some habitats, we stop people cutting down trees and hunting animals. The organisms in those habitats will continue to survive.

We also create new habitats and repair damaged habitats. One way of doing this is by planting trees in deforested areas.

However, it is up to us all as individuals to care for the habitats around us.

 21 Think of some ways that you could help the environment and explain how these ways will help.

Key facts:

✔ Increasing numbers of humans need more resources.

✔ Humans damage habitats by removing plants, hunting animals and causing pollution.

✔ Air pollutants from burning fossil fuels cause smog, acid rain and global warming.

✔ Many renewable energy resources cause less pollution than non-renewable resources.

✔ Renewable energy resources will not run out, unlike non-renewable resources.

✔ Humans help the environment by protecting and creating habitats, and reducing pollution.

Check your skills progress:

I can use evidence to make conclusions.

I can decide when to use a bar chart and when to use a line graph.

End of chapter review

Quick questions

1. A pond is an example of a:

   a resource   b environment

   c habitat   d physical factor   [1]

2. Some animals are adapted to seasonal changes by being able to:

   a hunt at night   b migrate

   c eat other animals   d sleep at night   [1]

3. An omnivore:

   a eats plants and animals   b eats only plants

   c eats only animals   d makes its own food   [1]

4. The last organism in a food chain is called the:

   a top predator   b primary consumer

   c producer   d carnivore   [1]

5. A gas produced by burning fossil fuels is:

   a methane   b oxygen

   c CFCs   d carbon dioxide   [1]

6. Sort these words into two lists: 'physical factors' and 'habitats'.

Arctic	cold	desert	dry
grassland	rainforest	wet	windy	[2]

7. Describe the physical factors that each of the following is adapted for.
   Give reasons for your choices.

   (a) Animal A stores water in its stomach.   [2]

   (b) Animal B has thick fur and small ears.   [2]

   (c) Plant C has very big leaves.   [2]

   (d) Animal D has very big eyes.   [2]

   (e) Animal E has large feet.   [2]

8. Some plants and animals have spines. How do spines help them survive? [1]

9. Scientists take samples from a habitat to find out what is living there. What method of sampling is best for these organisms?

   (a) Small plants. [1]

   (b) Animals that run along the ground. [1]

   (c) Small animals that live in tall grasses. [1]

10. Bioethanol is a type of biofuel.

    (a) What is a biofuel? [1]

    (b) Give *one* advantage of burning biofuels in cars rather than diesel. [1]

    (c) Give *one* disadvantage of biofuels. [1]

11. People often deforest jungles.

    (a) State what this means. [1]

    (b) Give *one* reason why people do this. [1]

    (c) State *two* resources needed by animals that deforestation removes. [1]

    (d) Describe *one* way in which people can help a deforested area. [1]

12. Burning fuels to produce electricity makes carbon dioxide.

    (a) Name the problem caused by too much carbon dioxide in the atmosphere. [1]

    (b) Describe a way of producing electricity that does not release this gas. [1]

    (c) Describe a disadvantage of this way of producing electricity. [1]

13. In winter, some bats migrate and other types of bat hibernate.

    (a) How does hibernation help the bats to survive? [1]

    (b) Describe *two* things that happens to a bat's body when it hibernates. [2]

    (c) How do bats prepare for hibernation? [1]

    (d) How does migration help bats to survive? [2]

    (e) What change in a physical factor causes bats to migrate or hibernate? [1]

Connect your understanding

14. Look at this food chain.

    grass → rabbit → fox

    In the food chain, identify a:

    (a) producer   [1]

    (b) consumer   [1]

    (c) top predator   [1]

    (d) herbivore   [1]

    (e) secondary consumer.   [1]

15. Owls hunt rats. Maize plants are food for rats.

    (a) Show these organisms in a food chain.   [2]

    (b) Where does the maize plant gets its energy from?   [1]

16. Scientists took samples in woodland in Turkmenistan. They wanted to find out about the small plants living among the trees. In the results table, the ticks show the plants found in each sample.

Type of plant	Sample number											
	1	2	3	4	5	6	7	8	9	10	11	12
A – nettle tree (*Celtis*)		✓		✓		✓		✓		✓	✓	
B – cherry (*Cerasus*)	✓						✓		✓		✓	
C – ephedra (*Ephedra*)			✓	✓		✓				✓		

    (a) Draw a bar chart to show the number of samples each plant was found in.   [4]

    (b) Which was the most common plant?   [1]

    (c) Why is this information not shown on a line graph?   [1]

17. The line graph shows changes in the thickness of the ozone layer above Antarctica.

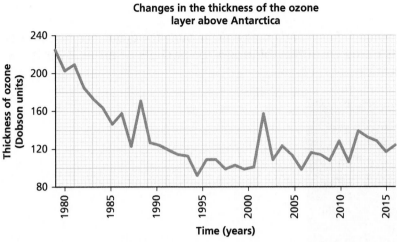

Changes in the thickness of the ozone layer above Antarctica

3.23 *Line graph to show changes in the thickness of the ozone layer.*

(a) Where is the ozone layer?   [1]

(b) Why is the ozone layer important for organisms?   [1]

(c) Describe the overall change in the thickness of the ozone layer from 1980 until the 1990s.   [1]

(d) Explain this change.   [1]

(e) Describe how the thickness of the ozone layer has changed since the 1990s. [1]

(f) Explain this change.   [1]

(g) One variable on the graph is time. What is the other variable?   [1]

18. A lion is a top predator but it provides food for other organisms. How does this happen? [2]

19. Bladderwrack is a seaweed. Seaweeds make their own food but are not plants. Bladderwrack gets its name from the 'bladders' or pockets of air all over it. Explain why it has this adaptation. [3]

3.24 *Bladderwrack.*

20. The information below is about organisms in the Southern Ocean. Use the information to draw a food chain.

- Weddell seals are predators of crabeater seals.
- Krill are tiny shrimp-like herbivores.
- Leopard seals are prey of killer whales.
- Diatoms are the producers in this food chain.
- Crabeater seals are the secondary consumers. [3]

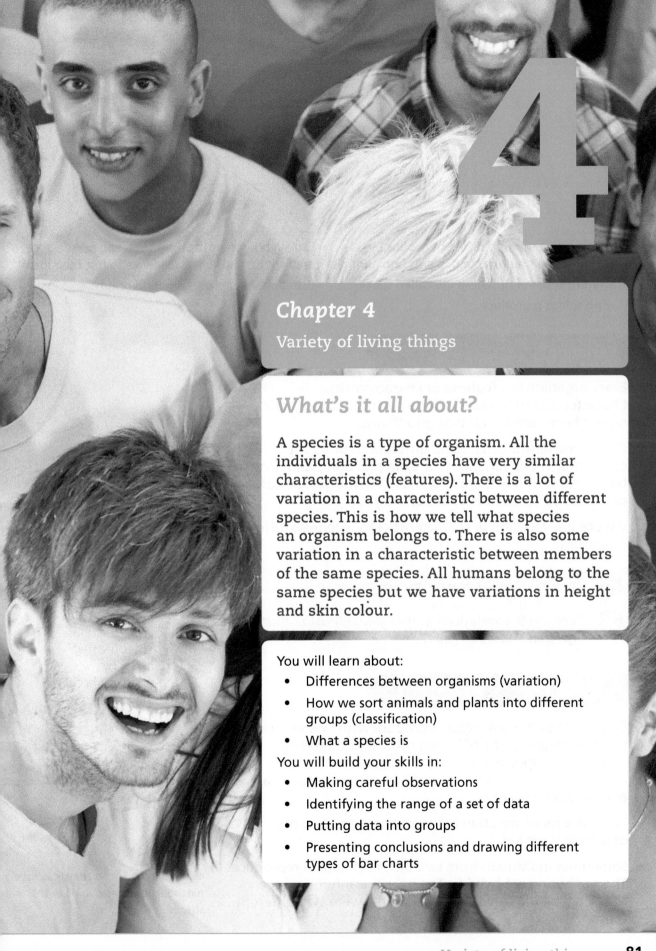

Chapter 4
Variety of living things

What's it all about?

A species is a type of organism. All the individuals in a species have very similar characteristics (features). There is a lot of variation in a characteristic between different species. This is how we tell what species an organism belongs to. There is also some variation in a characteristic between members of the same species. All humans belong to the same species but we have variations in height and skin colour.

You will learn about:
- Differences between organisms (variation)
- How we sort animals and plants into different groups (classification)
- What a species is

You will build your skills in:
- Making careful observations
- Identifying the range of a set of data
- Putting data into groups
- Presenting conclusions and drawing different types of bar charts

Species

Learning outcomes

- To recall what a species is
- To identify variation in the characteristics of different species
- To recall reasons why a species may become extinct

Starting point

You should know that...	You should be able to...
We put organisms into different groups by looking at their features (e.g. plants and animals)	Explain how scientists think up ideas, make predictions, and collect evidence to test their ideas
Different types of organism are called species	Use tables, bar charts and line graphs
Living things reproduce	

Variation

Every organism has features or characteristics. Characteristics of humans include having two legs, hair on our heads and hands that grip things.

One characteristic often looks different in different organisms. This is variation. For example, tigers and lions both have fur. They share that characteristic but the colour of the fur varies between lions and tigers.

We use the variation of characteristics to recognise different organisms.

1 Describe *three* characteristics of tigers.

2 Describe the variations in *two* characteristics that tigers and humans share.

Species

Organisms of the same type reproduce with one another to have offspring. The offspring grow and then they reproduce. A species is a group of organisms that can reproduce with one another and have offspring that can also produce offspring.

There is a lot of variation between the characteristics of different species.

Sometimes individuals from two different species reproduce. Their offspring are **hybrids**. The hybrids cannot produce offspring, which shows that the parents were two different species.

Key term

hybrid: offspring produced by reproduction between two different species.

4.1 *Characteristics of tigers include having legs and furry skin with stripes.*

3 **a)** State the names of the species in figures 4.1 and 4.2.

    b) Describe the variations in *two* characteristics of these species.

4 Horses and donkeys can reproduce. Their offspring are mules.

    a) Why is a mule a hybrid?

    b) What is a mule *not* able to do that its parents are able to do?

5 Suggest a characteristic of the liger in figure 4.3 that comes from its:

    a) father

    b) mother.

6 Suggest why we only find ligers in zoos (and not in the wild).

grandmother | mother (offspring of the grandmother)

daughter (offspring of the mother)

4.2 *Lions have offspring, which then also have offspring.*

Key term

extinct (life forms): does not exist any more.

4.3 *A liger is a hybrid. Its parents were a male lion and a female tiger. Ligers cannot have offspring.*

Extinction

Many of the characteristics of a species are adaptations that help it survive. There is variation between different species because they are adapted to different habitats.

Sometimes the environment of a habitat changes, and the characteristics of a species no longer help it survive. This can cause a species to become **extinct**.

An example is the woolly mammoth. About 10 000 years ago, there were many woolly mammoths in the Arctic. Then the temperature started to rise. Their hairy coats made the mammoths too hot and the plants that they ate could not survive in the higher temperatures. Woolly mammoths become extinct about 4000 years ago.

ears are quite large, to help it lose heat

tusks to dig for water and scrape off tree bark to eat

trunk to reach food and water (from in the trees and on the ground)

4.4 *Asian elephants have adaptations to survive in hot, humid rainforests.*

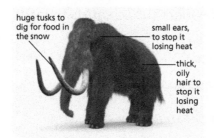

huge tusks to dig for food in the snow

small ears, to stop it losing heat

thick, oily hair to stop it losing heat

4.5 *Woolly mammoths had adaptations to survive extreme cold. This species is now extinct.*

4.6 *African savanna elephants live in hot habitats where there is not much shade.*

7 Dodo birds are extinct. What does this mean?

8 **a)** Figures 4.4, 4.5 and 4.6 show different species. What does this mean?

b) Describe the variation of the ears in the three species.

c) Explain how the variation in this characteristic helps survival.

d) Describe the variation in *two* more characteristics of these species.

e) Explain *two* reasons why woolly mammoths became extinct.

Scientists go on expeditions to look for species that may be extinct. If they do not find the organisms, this is evidence to support the idea that the species is extinct.

There are many reasons why a species may become extinct. These include:

- changes in physical factors
- habitat destruction
- pollution
- killing by people and animals
- other organisms using the resources.

9 **a)** Why were silversword plants in danger of becoming extinct in Hawaii?

b) Suggest what people did to stop the plants becoming extinct.

10 In 2006, scientists went to look for Yangtze river dolphins in the Yangtze River in China. They did not find any.

a) What scientific question were the scientists trying to answer?

b) What was their evidence?

c) What idea does this evidence support?

11 How can deforestation cause an animal to become extinct?

4.7 People brought sheep to Hawaii, and these sheep started to eat silversword plants. These plants were then in danger of becoming extinct.

Activity 4.1: Investigating extinction

Why are species becoming extinct?

A1 Scientists think that each of the following species may be extinct. The evidence to support these ideas is in brackets. Discover the reasons why each organism may be extinct. Use different books and the internet for your research.

- Christmas Island shrew (last seen in 1985)
- De Winton's golden mole (last seen in 1937)
- Dinagat Island cloud rat (last seen in 2012)
- Kouprey (last seen in 1988)
- Malabar civet (last seen in 1987)
- Telefomin cuscus (last seen in 1997)
- Yangtze river dolphin (last seen in 2002)

A2 Present your information as a table.

Biodiversity hotspots

'Biodiversity hotspots' are areas where many different species live. Scientists all over the world are counting the numbers of different species in different habitats to find biodiversity hotspots. They then protect these areas. For example, hunting may be banned. Protecting biodiversity hotspots should reduce the number of extinctions.

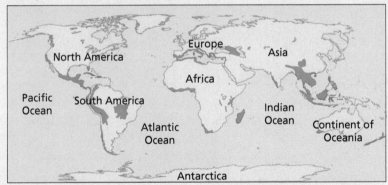

4.8 This map shows biodiversity hotspots on different continents.

12 **a)** What is a biodiversity hotspot?

   b) Explain why scientists want to protect biodiversity hotspots more than other areas.

13 **a)** Identify *one* continent that does not have a biodiversity hotspot.

   b) Suggest a reason why it does not have a biodiversity hotspot.

Check your skills progress:

I can find and use information from different sources.

I can identify scientific questions.

I can identify the evidence used to make conclusions.

I can present information using tables.

Classification

Learning outcomes
- To classify organisms as plants and animals
- To classify plants and animals into smaller groups

Starting point

You should know that...	You should be able to...
A species is a type of organism, and how we define a species	Find information using different sources
Different species have different adaptations to survive in their habitats	

Scientists put organisms into groups. This is classification. To classify organisms, scientists look at the variation of characteristics.

Kingdoms are the biggest groups of organisms. Examples include the **plant kingdom** and the **animal kingdom**. The table shows the characteristics used to classify plants and animals.

Kingdoms	
Plants...	**Animals...**
make their own food	eat other organisms
contain many cells	contain many cells
have cells with a cell wall made of a substance called cellulose.	have cells without cell walls
	are able to move their bodies from place to place.

1 Which kingdom do these organisms belong to?

a) cat

b) pine tree

c) lion

d) grass

e) elephant

2 State *one* characteristic that is the same in plants and animals.

There are other kingdoms too. The fungus kingdom contains mushrooms and yeasts. Another kingdom (called the 'prokaryote kingdom') contains bacteria. All bacteria are single cells, which do not have nuclei. The cells of organisms in all other kingdoms have nuclei.

Key terms

animal kingdom: kingdom that contains organisms that are made of more than one cell and are able to move their bodies from place to place.

kingdom: the biggest of the groups that scientists use to classify organisms.

plant kingdom: kingdom that contains organisms that are made of more than one cell and make their own food.

3 Fungi, such as mushrooms, feed on dead materials. Their cell walls contain a substance called chitin. Explain why fungi are in their own kingdom.

4 a) Yeasts are fungi with only one cell. How can you tell if a cell is a yeast or a bacterium?

b) What piece of equipment do you need to examine these cells?

Vertebrates and invertebrates

Scientists think that there are over 7 million different animal species. We can split the animal kingdom into smaller groups.

Classification is easier if groups have clear differences that are easy to see. For example, animals are **vertebrates** or **invertebrates**. A vertebrate has a hard skeleton inside it, including a 'backbone' made of bones called vertebrae. Invertebrates do not have skeletons but many of them have hard outer coverings.

Key terms

invertebrate: animal without a skeleton inside it and without a 'backbone'.

vertebrate: animal with a skeleton inside it, including a 'backbone'.

4.9 *Some animals.*

5 a) Look at figure 4.9. Are the animals vertebrates or invertebrates?

b) State the difference between vertebrates and invertebrates.

Vertebrates

Scientists divide vertebrates into five groups. Figure 4.10 shows these groups.

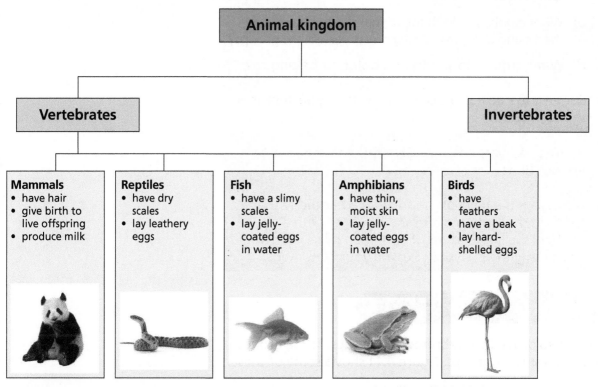

4.10 *There are five groups of vertebrates.*

6 Describe *one* characteristic of birds and reptiles that is:

   a) similar

   b) different.

7 a) What are animals with skeletons inside them called?

   b) List the groups into which scientists classify these animals.

   c) Name *two* animals from your area that belong to each group.

8 Which group of animals:

   a) has a backbone and feathers

   b) produces milk?

9 Which groups of animals:

   a) lay eggs

   b) have scales on their skin?

Key terms

amphibian: vertebrate with moist skin. It lays jelly-coated eggs in water.

bird: vertebrate with feathers. It lays eggs with hard shells.

fish: vertebrate with slimy scales. It lays jelly-coated eggs in water.

mammal: vertebrate with hair. It gives birth to live offspring.

reptile: vertebrate with dry scales. It lays eggs with a leathery coat.

10 Sharks and dolphins look similar.

   a) Describe *two* ways in which sharks and dolphins look similar.

   b) What would you look for to make sure you classify sharks and dolphins in the correct vertebrate groups?

   c) What groups do sharks and dolphins belong to?

11 Suggest the best characteristic to classify vertebrates in their groups.

12 Parrots and bats both fly. Give *two* reasons why bats are not birds.

13 A platypus is a hairy organism with a backbone that swims in rivers in Australia and lays eggs with a leathery shell. It feeds its offspring on milk. How would you classify the platypus? Explain your reasoning.

Invertebrates

Scientists also divide invertebrates into groups. These groups contain even smaller groups. Figure 4.11 shows some of these.

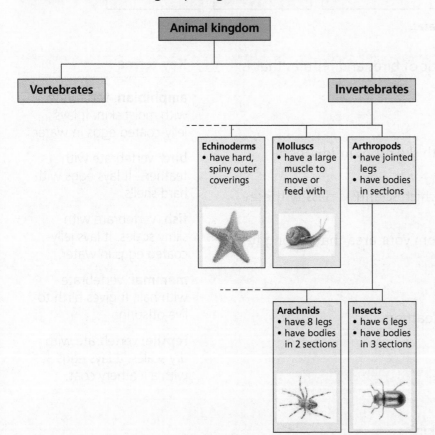

Key terms

arachnid: arthropod with eight legs and a body in two sections.

arthropod: invertebrate with jointed legs and a body in sections.

echinoderm: invertebrate with a hard, spiny outer covering.

insect: arthropod with six legs and a body in three sections.

mollusc: invertebrate with a large muscle that it uses to move or feed.

4.11 *There are many groups of invertebrates.*

14 **a)** Describe the characteristics of all the organisms in the arthropod group.

b) Name the groups that arthropods are divided into.

15 Describe *one* characteristic of arachnids and insects that is:

a) similar

b) different.

16 Look at figure 4.9 on page 88. List *all* the groups that these organisms belong to:

a) sea urchin

b) scorpion

c) fly.

17 A tick is a small blood-sucking invertebrate with eight legs.

a) Explain which group contains other animals with the most similar characteristics.

b) Give *one* other characteristic you would expect a tick to have.

Key terms

conifer: plant with needle-shaped leaves. It produces cones.

fern: plant that does not produce flowers or cones but has roots.

flowering plant: type of plant that produces flowers.

moss: plant with small, thin leaves. It does not have roots.

Plant groups

Scientists also divide the plant kingdom into groups. Figure 4.12 shows some of these.

```
                    Plant kingdom
```

Flowering plants
- have large, flat leaves
- have flowers
- have roots

Conifers
- have needle-shaped leaves
- have cones
- have roots

Ferns
- do not have cones or flowers
- have roots

Mosses
- have small, thin leaves
- do not have roots

4.12 *There are four main groups of plants.*

18 Describe *one* characteristic of mosses and flowering plants that is:

a) similar

b) different.

19 Which group of plants produces cones?

20 What variation is there in the leaves of conifers and flowering plants?

21 Name *two* flowering plants that live in your area.

Development of classification systems

People have classified organisms for thousands of years. 3500-year-old wall paintings provide evidence that the Ancient Egyptians classified plants. About 2300 years ago, Aristotle (an Ancient Greek thinker) wrote about plant classification. Indian writings from 1500 years ago also contain a classification system for plants.

A Swedish scientist, Carl Linnaeus, developed our modern classification system in the eighteenth century. He classified thousands of organisms into the groups that we use today. He also gave each species a two-word Latin name. Scientists all around the world still use this naming system, and agree on a single two-word scientific name for each species.

Biodiversity hotspots contain many undiscovered species. When scientists find a new organism, they use classification to work out what sort of organism it is.

4.13 *This animal was discovered recently in Vietnam. Its scientific name is* Cnemaspis psychedelica.

Activity 4.2: Classifying rare organisms

What rare organisms live in your country?

A1 Use different books and the internet to find the names of *ten rare* organisms in your country. Include at least *one* plant and at least *one* animal.

A2 Present your information as a table, showing the name of each organism and how it is classified. (You could try to use the scientific names for the organisms!)

22 Look at figure 4.13. Which group of vertebrates does this species belong to? Give a reason for your choice.

23 People give common names to the species living around them (such as 'blackbird', and 'green lizard'). Suggest why scientists use scientific names.

Key facts:

✔ Scientists classify organisms in groups by looking at their characteristics.

✔ The largest classification groups are kingdoms.

✔ The animal kingdom contains five groups of vertebrates – mammals, reptiles, fish, amphibians and birds.

✔ The animal kingdom contains many groups of invertebrates, such as arachnids and insects.

✔ The plant kingdom contains groups such as flowering plants and conifers.

Check your skills progress:

I can use characteristics to put items into groups.

I can find and use information from different sources.

Types of variation

Learning outcomes
• To identify discontinuous and continuous variation
• To choose charts and graphs to display different sorts of information

Starting point

You should know that...	You should be able to...
Scientists classify organisms by putting them into groups, based on their characteristics	Use tables, bar charts and line graphs
Characteristics vary between one organism and another	

Proboscis monkeys and orangutans live on the island of Borneo. They share many characteristics, such as having a nose and ears. However, there are big variations in those characteristics because they are different species.

There is much less variation in a characteristic between members of the same species. For example, all humans have the same type of skin but it varies in colour.

4.14 *A proboscis monkey.*

1. Look at figures 4.14 and 4.15.

   a) Both proboscis monkeys and orangutans have noses. Describe the variation in this characteristic between the two species.

   b) Choose *one* other characteristic of the animals. Describe how it varies.

2. Look at someone nearby. Describe the variation in *two* characteristics that you and that person both have.

4.15 *An orangutan.*

Discontinuous variation

Variation with a distinct set of options is **discontinuous variation**. For example, some people are able to roll their tongues and others are not. There are only two possible options in the characteristic of tongue rolling – 'able to roll' and 'not able to roll'.

4.16 *There is variation in the colour of human skin.*

Key term

discontinuous variation: variation that has a distinct set of options or categories.

Another example is the number of peas in a pea pod, which is a **whole number**. Some peas are smaller than others, but fractions of peas do not grow.

We show discontinuous variation on a bar chart, with gaps between the bars.

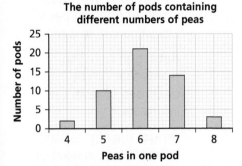

4.17 Scientists counted the number of peas in 50 different pea pods.

4.17 Can you roll your tongue?

3 Look at the bar chart.

a) How many pods contained six peas?

b) What was the least common number of peas in one pod?

4 The table shows the number of nails on the front feet of elephant species.

Elephant species	Number of nails on front feet
African forest elephant	5
African savanna elephant	4
Asian elephant	5

a) Explain why this data shows discontinuous variation.

b) Present the data using a suitable chart or graph.

Activity 4.3: Investigating discontinuous variation

Find out if your friends or family are able to roll their tongues.

A1 Draw a table to show the number of people who are able to roll their tongues and the number who are not able.

A2 Present your data using a suitable chart or graph.

A3 Make a conclusion.

A set of **data** has a **range**. To describe a range we find the highest and lowest values. In the tables below, the top value is 104 mm and the bottom value is 80 mm. We say that the range is 'from 80 to 104 mm'.

Variation that may have any value in a range is **continuous variation**. For example, very few people in your class have exactly the same height. Someone's height may have any value within a range.

To show continuous variation on a bar chart, we put the data into groups of smaller ranges. These ranges must not overlap and must all be the same size. Here is an example using pea pod length.

- *Group your data:* Look at the data showing pea pod lengths. Groups could be: 80–84 mm, 85–89 mm, 90–95 mm and so on. We do not include groups outside the range of the data.

Key terms

continuous variation: variation that can have any value within a range.

data: numbers and words that can be organised to give information.

range: the highest and lowest values in a set of data.

Lengths of some pea pods (mm)

80 99 104 84 96 94 97 98 89 90 88 91 93 96 87 88 90

- *Draw a tally chart:* We write the groups into a table called a tally chart.

Grouped lengths of pea pods (mm)	Tally	Total
80–84		
85–89		
90–94		
95–99		
100–104		

- *Complete the tally:* One by one, we cross out each value and put a mark in the 'tally column' to show its group.

Lengths of some pea pods (mm)

80 91 104 84 93 92 94 91 89 90 88 91 98 96 87 88 96 86 90 99

Grouped lengths of pea pods (mm)	Tally	Total
80–84	//	
85–89		
90–94	////	
95–99		
100–104	/	

When you have four tally marks and add a fifth, you draw it through the other four. So //// means '5'.

- *Complete the totals:* We add up each tally and write in the totals.

Grouped lengths of pea pods (mm)	Tally	Total
80–84	//	2
85–89	////	5
90–94	//// ///	8
95–99	////	4
100–104	/	1

- *Draw a bar chart without gaps:* We show the grouped data on a bar chart. The data is continuous so we do not have gaps between the bars.

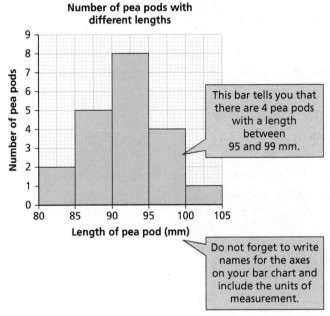

Number of pea pods with different lengths

This bar tells you that there are 4 pea pods with a length between 95 and 99 mm.

Do not forget to write names for the axes on your bar chart and include the units of measurement.

4.18 *Bar charts with grouped continuous data do not have gaps between the bars.*

Activity 4.4: Investigating continuous variation

- Take off your shoes and stand against a wall or a board.
- Ask someone to put a pencil on top of your head and make a mark.
- Use a tape measure to measure the height of the mark. This is your height.

  A1 Collect height measurements from others in your class.

  A2 State the range of the measurements.

  A3 Design groups for the measurements.

  A4 Draw a tally chart for your data.

  A5 Present your data using a bar chart.

  A6 Make a conclusion about which heights are most common.

4.19

5 State whether each of the following describes continuous or discontinuous variation:

a) heights of trees

b) lengths of leaves

c) having an earring

d) number of times your heart beats in 10 seconds

e) the sizes of cakes.

6 Look at the data in the table below, which shows some human heights.

> 1.76 m 1.70 m 1.56 m 1.87 m 1.60 m 1.67 m 1.75 m
> 1.83 m 1.61 m 1.84 m 1.82 m 1.77 m 1.72 m 1.57 m

a) Is this continuous or discontinuous variation?

b) What is the range of the data?

c) Draw a tally chart from the data. Use these groups:

   1.50–1.59 m, 1.60–1.69 m, 1.70–1.79 m, 1.80–1.89 m

d) Present this data using a suitable chart or graph.

7 Some ID cards contain information about variation. When a person uses the card, a computer checks the variations in the real person with the information stored in the card.

Is it better to use continuous or discontinuous variation about humans for ID cards?
Explain your reasoning.

Quick questions

1. A characteristic that all mammals have is:

   a hair   **b** laying eggs

   c wings   **d** scales   [1]

2. Grouping organisms with similar characteristics is:

   a classification   **b** organising

   c boxing   **d** dividing   [1]

3. An animal is placed in the vertebrate group if it has:

   a scales   **b** blood

   c a backbone   **d** hair   [1]

4. Plants that have needle-shaped leaves are in the group of:

   a fungi   **b** conifers

   c flowering plants   **d** ferns   [1]

5. Variation that has a limited range of options is:

   a characteristic   **b** continuous

   c discounted   **d** discontinuous   [1]

6. Read these word meanings. State the words that match each meaning.

   (a) A group of organisms that can reproduce with one another and have
   offspring that can also produce offspring.   [1]

   (b) The offspring produced when two different species reproduce.   [1]

   (c) When a type of organism stops existing.   [1]

   (d) Data used to show that an idea is correct or incorrect.   [1]

7. State whether each of the following describes continuous or discontinuous
   variation in humans:

   (a) length of hair   [1]

   (b) wearing glasses   [1]

   (c) having a scar   [1]

    (d) weight [1]

    (e) height [1]

    (f) arm length. [1]

8. Name *one* kingdom that contains organisms that:

    (a) have cell walls made of cellulose [1]

    (b) move their bodies from place to place [1]

    (c) make their own food. [1]

9. What characteristic of flowering plants does no other plant group have? [1]

Connect your understanding

10. Look at figures 4.20 and 4.21, which both show vertebrates.

4.20 *Brahman cattle are common around the world.*

4.21 *Yaks live mainly in the Himalayas.*

    (a) Which kingdom do vertebrates belong to? [1]

    (b) What group of vertebrates do both these organisms belong to? [1]

    (c) Describe *three* characteristics that all members of this group of vertebrates have, but that other groups of vertebrates do not have. [3]

    (d) Choose *two* characteristics that vary between the two organisms and describe the variation. [2]

11. The lli pika is an animal that looks a bit like a small rabbit. It is a herbivore and lives high in the Tianshan mountains in China. The numbers of the animal are decreasing. Scientists think that it might become extinct.

    (a) Describe *one* adaptation you think the animal has. Explain your reasoning. [2]

    (b) Suggest *one* reason why the lli pike might become extinct. [1]

12. Scientists classify human blood into four different 'blood groups'. These are: A, B, AB and O. Scientists in Saudi Arabia looked at the blood groups of 100 people. The table shows their results.

Blood group	Number of people
A	26
B	18
AB	4
O	52

(a) Is variation in blood group continuous or discontinuous? Explain your choice. [2]

(b) Present the data using a suitable chart or graph. [2]

13. The bar chart in figure 4.22 shows the results of the men's discus competition at the Rio Olympics in Brazil. The competition involves throwing a disc as far as possible.

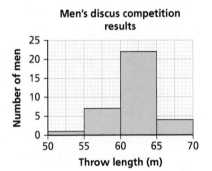

Men's discus competition results

(a) Is variation in the bar chart continuous or discontinuous? Explain your choice. [2]

(b) How many men threw a distance between 55 and 59 m? [1]

(c) What was the most common range of distances thrown? [1]

4.22

14. Look at figure 4.23.

(a) Classify organisms W, X, Y and Z. Explain how you make your choices. [8]

(b) The characteristics that you have used to make your choices vary. Is this variation continuous or discontinuous? Explain your answer. [2]

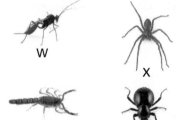

W

X

Y

Z

4.23

15. The tally chart shows some lengths of beans. Some parts of it are missing.

Grouped lengths of beans (mm)	Tally	Total
70–74	//	
75–79		5
80–84	### /	
	////	
90–94	///	

Copy and complete the chart. [6]

16. List *two* groups of vertebrates that:

    (a) lay jelly-coated eggs in water   **[2]**

    (b) have scales.   **[2]**

Challenge question

17. A student measured the widths of some leaves from a large garden plant. Their results are shown below.

6.8	7.0	5.5	5.6	7.7
6.7	7.2	5.9	5.5	5.8
6.2	5.9	6.1	6.2	7.6
6.0	6.7	6.9	5.8	7.1
5.7	6.8	6.4	6.3	7.0

    (a) The student forgot to write down the units for measuring the leaves. Suggest what the units are.   **[1]**

    (b) What is the range of the data?   **[1]**

    (c) Draw a tally chart to display this data.   **[2]**

    (d) Use the data from your tally chart to draw a bar chart.   **[2]**

1. (a) (i) The table shows the parts of a cell. The functions are not in the correct order. Copy the table and put the functions of each part in the correct order.   [1]

Part	Function
cell membrane	makes new substances
chloroplast	controls the cell
cytoplasm	makes food
nucleus	controls what enters and leaves the cell

(ii) Explain how you know that this is a plant cell.   [1]

(b) The diagram shows some muscles in the leg.

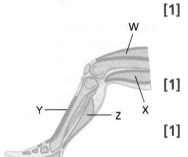

(i) Give the reason why many muscles are found in antagonistic pairs.   [1]

(ii) State the letter of the muscle that contracts to point the toes out straight.   [1]

(c) Four different trees are planted in the same area. Their heights are measured every year. The graph shows this data.

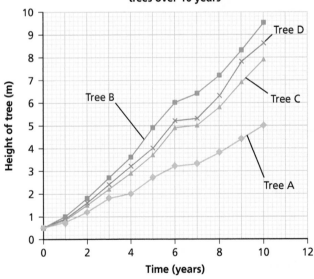

The growth of four different trees over 10 years

    (i) Which life process is shown in the graph? [1]

    (ii) How tall was tree A after four years? [1]

(d) Copy and complete this table to show the functions of some different organ systems and *two* organs found in each. [1]

Organ system	Function	Two organs
		heart, blood vessels
	to digest food and absorb it into the blood	

(e) Red blood cells have an indented shape. Explain how this is an adaptation for their function. [1]

2. (a) In the seventeenth century, many people thought that meat made maggots. Francesco Redi did not believe this. He did an experiment to get evidence to support his idea that maggots in meat came from flies. The drawing shows part of the experiment.

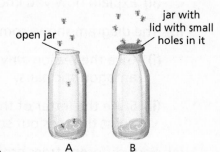

    (i) Make a prediction about what will happen in jars A and B. Explain your predictions. [1]

In the nineteenth century, a scientist showed that something from the air made soup spoil. The drawing shows part of one of his experiments.

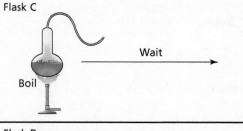

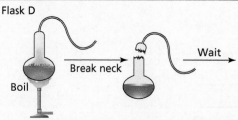

    (ii) What was the name of this scientist? [1]

    (iii) Make a prediction about what will happen in flasks C and D. Explain your predictions. [1]

(b) Each year scientists make a new vaccine for flu (influenza).

    (i) What is a vaccine? [1]

    (ii) Explain why the scientists making the vaccine wear face masks. [1]

(c) A student made batches of bread dough, with different amounts of sugar in each batch. The student put each dough in a measuring cylinder and left them all for 40 minutes. The table shows the results.

Batch	Mass of sugar added (g)	Increase in the height of the dough after 40 minutes (cm)
1	0	3
2	2	4
3	4	10
4	6	15
5	8	19

    (i) Explain why the dough rises. [1]

    (ii) State *two* variables that should be kept the same. [1]

    (iii) Use the evidence in the table to make a conclusion. [1]

3. (a) The drawings show an Arctic fox, a desert fox and a red fox.

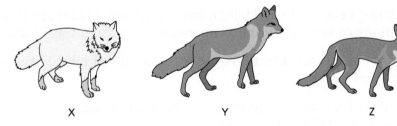

thick white fur   medium thickness red fur   thin brown fur

X   Y   Z

    (i) Which is the Arctic fox? [1]

    (ii) Explain *two* ways in which the desert fox is adapted to its habitat. [1]

(b) Arctic hares feed on saxifrage plants. Ermine are prey of snowy owls. Ermine are predators of Arctic hares.

(i) Show these organisms in a food chain. [1]

(ii) In the food chain, give *one* organism that is a top predator, and *one* that is a primary consumer. [1]

(c) The bar chart shows the number of bird deaths caused by wind turbines in the USA.

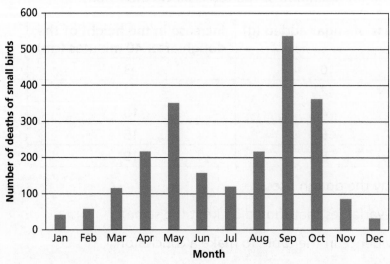

(i) Wind turbines are a renewable source of energy. What does renewable mean? [1]

(ii) Apart from being renewable and killing birds, give *one* advantage and *one* disadvantage of wind turbines compared with burning fossil fuels. [1]

(iii) Suggest an explanation for why there are more deaths of birds in some months compared to others. [1]

4. (a) A group of organisms that can reproduce with one another and have offspring that can also produce offspring is: [1]

　　a a hybrid　**b** a kingdom　**c** an echinoderm　**d** a species

(b) A small creature with thick white fur lives only in the high areas of Mount Lewis in Australia. Its name is the white lemuroid ringtail possum. Some scientists think that global warming will make it extinct.

(i) This animal is a vertebrate. What type of vertebrate is it? Explain your reasoning. [1]

(ii) Suggest an explanation for why this animal may become extinct. [1]

(c) A student measured the lengths of some leaves on one type of plant. The student measured the leaves in centimetres with a ruler.

10.8	11.0	10.5	12.6	11.7
12.7	10.2	12.9	12.5	12.0
12.2	11.9	10.1	11.2	12.6
11.0	11.7	11.9	12.8	12.1
11.7	10.8	11.4	11.3	12.0

(i) What type of variation does this data show? [1]

(ii) Copy and complete this tally chart for this data. [1]

Grouped lengths of leaves (cm)	Tally	Total
10.1–10.5		
10.6–11.0		
11.1–11.5		
11.6–12.0		
12.1–12.5		
12.6–13.0		

(iii) What type of graph or chart would you use to plot the data from your tally chart? [1]

(d) Copy and complete these sentences.

Mushrooms and yeasts belong to the _____ kingdom.

Bacteria belong to the 'prokaryote kingdom'. Unlike organisms in all other kingdoms, the cells of prokaryotes do not contain

_____. [1]

[total 30 marks]

Chemistry

Chapter 5
Properties of matter and materials

What's it all about?

Engineers can build very tall towers, like the Burj Khalifa tower in Dubai, because they choose materials, like steel and concrete, which are strong. Steel is strong because the tiny particles in a metal are held together strongly. These forces between particles also make steel hard to melt. To change steel into a liquid the temperature must be over 1000 °C.

You will learn about:
- How the arrangement and movement of tiny particles give solids, liquids and gases different properties
- What happens to these particles during changes of state, including melting and boiling
- The different physical properties of materials, and why materials are chosen for different uses
- The differences between metals and non-metals, and how to test their properties

You will build your skills in:
- Choosing the right equipment to investigate a question
- Drawing line graphs and recognising results that fit a pattern
- Writing predictions, and using results from an investigation to decide if your prediction is correct

The states of matter

Learning outcomes
- To use particle theory to explain the properties of solids, liquids and gases
- To make careful observations

Starting point

You should know that...	You should be able to...
Substances can be solids, liquids or gases	Sort (group) materials using their properties
	Make a prediction and say if evidence supports a prediction

Solids, liquids and gases

Substances exist as solids, liquids and gases. These are the three **states of matter**.

What solids, liquids and gases are in this photograph?

5.1 *The mountain, lake and the air in this picture are made of different substances.*

Activity 5.1: The properties of states of matter

Water can exist in all three states.

You have three balloons containing water in each state.

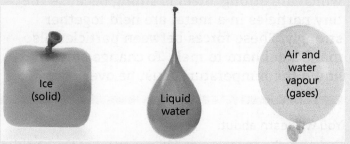

Ice (solid)

Liquid water

Air and water vapour (gases)

5.2 *Your breath contains water **vapour** because the inside of your lungs is wet. So, when you blow into a balloon, the balloon contains air and water vapour.*

Observe each balloon.

Use these tests to compare the **physical properties** of solids, liquids and gases.

For each balloon:

- can you compress (squash) it?

- can the contents flow (move from place to place)?

Your teacher will burst the balloons. Make a prediction: will the substance inside stay the same shape or change?

Key terms

physical properties: the properties of an object that can be observed and measured.

state of matter: the three forms that a substance can exist in: solid, liquid and gas.

vapour: liquid that has evaporated to form a gas.

Properties of solids, liquids and gases

Solids have a fixed shape and **volume**. They cannot flow or be compressed.

Liquids have a fixed volume but not a fixed shape. They can flow but cannot be compressed.

Gases do not have a fixed shape or volume. They can flow and be compressed.

1. For each statement, identify the state of matter.
   a) The state of matter that can flow but cannot be compressed.
   b) Wood and ice are always in this state.
   c) The state of matter that *cannot* flow.
   d) The only state of matter that can be compressed.

2. Tariq reads in a book that *water is a liquid*. Explain why this is not always true.

3. Sarah and Zoha are talking about the states of matter.

   Sarah says that sand is a solid.

   Zoha says that sand must be a liquid because it can flow through your hands.

   Who is correct? Give a reason for your answer.

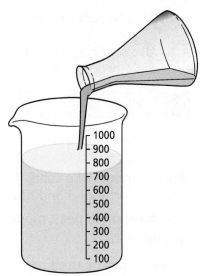

5.3 *You can pour a liquid from one container into another because liquids can flow. The volume will be the same in each container but the liquid's shape will change.*

Particle theory

All solids, liquids and gases are made of tiny particles such as atoms or molecules. You cannot see these particles because they are much too small.

The properties of solids, liquids and gases are different because their particles are arranged differently. This **model** is called **particle theory**. In science, models help us understand something we cannot actually see. When we use a model, we must remember it is just a way to think about something. There will be some differences to the real thing.

Key terms

model: simple way of showing or explaining a complicated object or idea.

particle theory: model that describes how particles are arranged differently in solids, liquids and gases.

solid

5.4 *In a solid the particles are all touching.*
This explains why you cannot compress a solid.
The particles are held together by strong forces.
This explains why the shape of a solid does not change. It is fixed.
The particles are vibrating (shaking from side to side) but they cannot move around.
This explains why solids cannot flow.

liquid

5.5 *The particles in a liquid are all touching.*
This explains why you cannot compress a liquid.
The particles are moving; they can move past one another.
This explains why liquids can flow and why they take the shape of their container.

gas

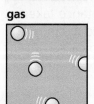

5.6 *The particles in a gas are far apart.*
This explains why you can compress a gas; you can push the particles together.
The particles are moving around very fast in all directions.
This explains why gases can flow and fill their container.

5.7 *When a gas is compressed the particles are pushed closer together.*

4 What state of matter is being described?

   a) The particles are far apart.

   b) The particles are not moving around.

   c) The particles are all touching but they are moving around.

5 Use particle theory to explain why you can compress a gas, but you *cannot* compress a solid or a liquid.

6 A diamond is hard and cannot be scratched easily. Use particle theory to suggest why.

Activity 5.2: Building a model

Build a model of a solid, liquid and a gas. Your model should show the particles, how they are arranged and how they move.

You have dried peas, modelling clay, and a plastic bottle with a lid to build your model.

Think about:

- What can be used to represent the particles?
- The particles in a solid are held together by strong forces. How can you model this?
- Use the bottle as a container to put your model liquid or model gas into. How can you model how the particles are arranged and move in a liquid and a gas? Hint: you can move the bottle.

Key facts:

✔ Substances are made up of tiny particles.

✔ The three states of matter are solids, liquids and gases.

✔ Solids, liquids and gases have different properties because their particles are arranged differently. This model is called particle theory.

Check your skills progress:

I can make careful observations.

I can group substances as solids, liquids and gases.

I can use particle theory to explain the properties of solids, liquids and gases.

Changing state

Learning outcomes

- To use particle theory to explain what happens during changes in state
- To draw a line graph and describe the pattern

Starting point

You should know that...	You should be able to...
Melting occurs when a solid turns into a liquid and is the reverse of freezing	Measure temperature and time
The boiling point of water is 100 °C and the melting point of ice is 0 °C	Use results tables and line graphs
Evaporation occurs when a liquid turns into a gas	
Condensation occurs when a gas turns into a liquid and is the reverse of evaporation	

Changes in state

If you leave ice in a warm place, it changes from a solid into a liquid (water).

This change in state is called **melting**.

Changes in state are **reversible changes**. So, if you put the water from the melted ice into a freezer, it will freeze and change back into ice.

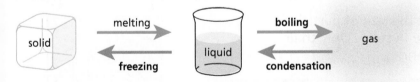

5.8 *Some changes in state.*

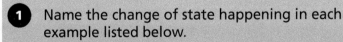

1 Name the change of state happening in each example listed below.

  a) A gold ring is heated until it turns into liquid gold.

  b) Drops of water form on a bathroom mirror.

  c) A chocolate bar is left in a warm kitchen.

  d) A layer of ice forms on top of a lake in winter.

Key terms

boiling: the change of state from liquid to gas.

condensation: the change of state from gas to liquid.

freezing: the change of state from liquid to solid.

melting: the change of state from solid to liquid.

reversible change: change in a substance that can be changed back again.

2 Three students were asked what would happen to the mass of an ice cube as it melted.

Who is correct?

Aisha: The mass will go down because liquids have less mass than solids.

Zain: The mass will stay the same because no particles in the ice have been lost or gained.

Adnan: The mass will go up because the water takes up more space than the ice.

5.9 *Steam is a gas formed when water boils. It is invisible. What you see above the kettle are tiny droplets of water formed when the steam condenses in the air.*

Explaining changes in state

When a change in state happens, the energy of the particles changes.

During melting and boiling, heating increases the energy of the particles in the material.

During freezing and condensation, the energy of the particles decreases.

You can use particle theory to explain what happens when a material changes state.

When a solid melts the particles vibrate faster. Some of the particles gain enough energy to move away from the others. Eventually, all the particles are moving faster and slightly further apart. The material has melted and is now a liquid.

When a liquid freezes the particles move around less. Some of the particles move even closer together. Eventually the particles cannot move around, only vibrate in a fixed position. The material has frozen and is now a solid.

It is important to remember that the size of particles does not change during a change in state. Only their arrangement and movement changes.

3 Describe what happens to the particles when a liquid is heated to form a gas.

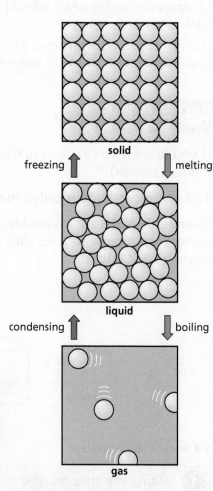

5.10 *The red arrows show heating. The particles move more, weakening the forces between them. The blue arrows show cooling. The particles move less, so the forces between them increase in strength.*

Melting and boiling points

The temperature at which a material melts is called its **melting point**. This is the same as its freezing point.

The temperature at which a material boils is called its **boiling point**.

This is the method to measure the melting point of a solid and the boiling point of a liquid.

A Place the material (a solid or liquid) into a suitable container. Use a thermometer to measure its temperature.

B Heat the material.

C Use a thermometer to measure the temperature every minute until the temperature stops changing. This is the melting or boiling point.

Key terms

boiling point: the temperature a substance boils at, and changes from a liquid into a gas.

melting point: the temperature a substance melts at, and changes from a solid into a liquid.

The temperature was taken at the start, before she started heating the water. This is 0 minutes.

The variable you change is written in the first column.

Time (minutes)	Temperature of water (°C)
0	21
1	45
2	67
3	84
4	93
5	98
6	100
7	100
8	100

Each column has a heading that includes the units of the measurement.

The variable you measure is written in the second column

The temperatures are written in order, with the lowest first.

Table 5.1 *Results table for measuring the boiling point of water.*

Activity 5.3 Measuring the boiling point of water

Aisha measured the boiling point of water.

She used a results table to show her results (table 5.1).

A1 Plot the results in the table as a line graph. Remember:

- The variable that was changed (time) goes on the horizontal axis.
- The variable that was measured (temperature) goes on the vertical axis.
- Label each axis. Use the column headings from the table.
- Plot each point as a cross.
- Join the crosses up.

A2 Describe the pattern in the graph by filling in these sentences:

Between the time of _____ minutes and _____ minutes the temperature increased from _____°C to _____°C.

At the time of _____ minutes the temperature stayed at _____°C and did not increase any more.

A3 What is the boiling point of water? How can you tell from the graph?

4 A scientist wanted to find the melting point of naphthalene.

They put some solid naphthalene into a test tube and heated it until it melted. Then they put the test tube into a beaker of ice to cool the naphthalene.

They measured the temperature of the naphthalene every minute for 18 minutes.

They drew this graph.

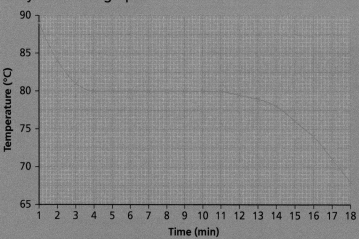

5.11 *A graph to show the temperature of liquid naphthalene as it cools.*

a) Describe the pattern in the results.

b) Describe what the scientist would have observed between 4 and 11 minutes.

c) Use the graph to find out the melting point of naphthalene.

What is the difference between boiling and evaporation?

Wet clothes dry because the water in them changes into a gas (water vapour).

But, the water has not boiled. Boiling only happens if the water reaches a temperature of 100 °C.

The water has evaporated.

Both boiling and **evaporating** are changes of state when a liquid turns into a gas.

Boiling only happens at the liquid's boiling point.

Evaporation can happen at any temperature below the boiling point.

Key term

evaporating: the change of state from liquid to gas that happens below the boiling point.

Using the particle model to explain evaporation

When a liquid evaporates some of the particles at the surface gain enough energy to move away from the other particles.

The particles spread out. They are now in a gas.

When a liquid boils all the particles in the liquid gain enough energy to move away from each other, so the liquid quickly changes into a gas. Bubbles of gas form throughout the liquid. They rise to the surface and then leave the liquid.

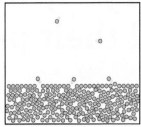

5.12 *During evaporation some of the particles in the liquid leave the liquid as a gas.*

5.13 *The bubbles in boiling water contain a gas (steam).*

Using evaporation

Traditional clay pot coolers keep food cool without electricity. They work because of evaporation.

There are two pots, separated by a layer of sand. Water is poured onto the sand.

As the water evaporates from the sand, it transfers energy away from inner pot, keeping the food cool.

Key facts:

✔ Heating makes particles move more.

✔ If the particles in a solid gain enough energy, the attractive forces between them cannot hold them together. The solid melts to become a liquid.

✔ If the particles in a liquid lose enough energy, the particles move less and the attractive forces between them can now hold them together. The liquid freezes to form a solid.

✔ Melting happens at the material's melting point.

✔ Boiling happens at its boiling point.

Check your skills progress:

I can draw a line graph to show how the temperature of a material changes as it is heated or cooled.

I can describe the pattern in a graph.

Everyday materials and their properties

Learning outcomes

- To describe the physical properties of materials
- To make a prediction and decide if evidence supports it
- To choose suitable apparatus to collect evidence to test an idea and use it correctly

Starting point

You should know that...	You should be able to...
Different materials have different properties such as hard, soft, shiny	Sort (group) materials using their properties
Materials are chosen for different uses because of their properties	Make a prediction and say if evidence supports a prediction
	Choose what evidence to collect to investigate a question
	Design a fair test

5.14 *Diamond drills are used to scratch designs into glass. This works because diamond is harder than glass.*

Physical properties

Materials have different physical properties. These are things that can be observed and measured. Comparing the results helps us to choose the best material to use.

Hard materials cannot easily be scratched or dented.

Strong materials need a large force to break them.

Safer building materials

Earthquakes often happen in Indonesia. The materials used to make bricks need to be strong so buildings do not collapse. The local people who make bricks are getting help from organisations who show them how to make stronger bricks that will help keep their community safe.

5.15 *The rope used to build rope bridges has to be very strong.*

Activity 5.4: Comparing hardness

Your teacher will give you some different materials.

A1 In a group, discuss how you could compare the hardness of the materials.

A2 Use your method to list the materials in order of how hard they are.

5.16 *Strong bricks can help protect people during earthquakes.*

Absorbent materials

Some materials are **absorbent**. This means liquids soak into them.

Other materials do not absorb liquids.

Activity 5.5: Which type of paper is the most absorbent?

You will be given some different types of paper. Your task is to investigate how absorbent each is.

Think about these questions:

A1 The type of paper is the variable you will change. What variable will you measure?

A2 What equipment will you need to do this?

A3 How will you use it to answer the question?

A4 What variables will you keep the same? These are your **control variables**.

A5 Which paper do you think will be the most absorbent? Why do you think this? This is your **prediction**.

A6 When you have collected your results use them to say which paper was the most absorbent.

A7 Was your prediction correct?

5.17 *A sponge is absorbent.*

> **Key terms**
>
> **absorbent**: soaks up liquids.
>
> **control variables**: variables that you keep the same during an investigation.
>
> **prediction**: what you think will happen in an investigation.

Flexible, brittle and malleable materials

Flexible materials are easy to bend and will not break when bent. Clothes made of cotton are flexible.

A **brittle** material will break when it is bent. Ceramic tiles are brittle.

Malleable materials can be made into a new shape. They can be squashed, flattened or bent and will stay in their new shape. Clay is malleable; it can be used to form many different shapes.

> **Key terms**
>
> **brittle**: breaks when bent.
>
> **flexible**: can be easily bent and will not break.
>
> **malleable**: can be formed into different shapes.

5.18 *A twig is brittle.*

5.19 *A rubber ball is flexible.*

5.20 *Bread dough is malleable.*

1. State if each of these materials is flexible, brittle or malleable:

    a) glass jug

    b) plastic drinking straw

    c) wooden pencil.

2. Name a material that has these physical properties:

    a) flexible and absorbent

    b) hard and brittle

    c) malleable and flexible.

Opaque and transparent materials

Glass is a good choice of material for windows because it allows light to pass through it. It is **transparent**.

Other materials are not transparent. They are **opaque**. Opaque materials do not let light pass through them. Wood is an example of an opaque material.

Some materials have different types that have different properties. For example, some types of plastic are transparent. Others are opaque.

3. Name an object in a kitchen that is:

    a) transparent

    b) opaque.

4. Glass is not always transparent. Explain why this is true.

5. Describe how you could test different plastics to see which is the most transparent.

Key terms
....

opaque: light cannot pass through it.

transparent: light can pass through it.

5.21 *The plastic used for the cups is transparent. The plastic straws are opaque.*

Key facts:

✔ Materials have physical properties such as absorbent, flexible, brittle, malleable, transparent, opaque.

✔ Materials are chosen for a particular use because of their properties.

Check your skills progress:

I can choose the correct equipment to collect evidence to answer a question.

I can write a prediction using scientific knowledge.

I can use evidence collected from an investigation to say whether a prediction is correct, or not.

Comparing metals and non-metals

Learning outcomes
- To decide if a material is a metal or non-metal based on its physical properties
- To make conclusions from evidence

Starting point

You should know that...	You should be able to...
Some materials are better conductors of electricity than others	Use results to draw conclusions
Some metals are good conductors of electricity and most other materials are not	

Metals and non-metals

All materials can be grouped as either a metal or a non-metal, using their physical properties.

Metals have useful physical properties, for example many are hard and strong. Examples of metals include gold, copper, iron and steel.

Metals are used to make many different objects.

A material that is not a metal is a non-metal. Examples of non-metals are plastic, glass, water and air.

Non-metals have different physical properties to metals. They are dull, not shiny, and often have low melting and boiling points.

5.22 *Steel is hard, strong and malleable. It is used to make bridges.*

Metals and non-metals in the Burj Khalifa tower

Most buildings in hot places do not have many windows. This is because heat, as well as light, passes through glass. The Burj Khalifa has 24 000 windows to let the people inside enjoy the views. The glass used has a very thin layer of metal on it. The metal lets the light pass through but heat is reflected. This keeps the building cool inside.

5.23 *The Burj Khalifa tower in Dubai is built from a non-metal called concrete. Steel rods are put inside the concrete to add strength.*

Testing the properties of metals and non-metals

Most metals have these physical properties: strong, hard, high melting and boiling point, shiny and malleable.

But, it is very difficult to decide if a material is a metal based on just these properties. For example, not all metals are hard. Lead is very soft and can also easily be bent by hand.

Also, some non-metals have these properties. Diamond is a non-metal that is very hard. There are some other properties that can be tested to decide if a material is a metal or non-metal.

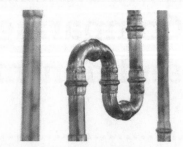

5.24 *Copper is malleable so can be bent to form water pipes.*

Activity 5.6: Metal or non-metal?

Your teacher will give you some objects. Your task is to decide if each object is made of a metal or a non-metal and place them into two groups.

Think about these questions:

- What physical properties can you observe?
- Can you decide if it is a metal or non-metal just by observing the material?

5.25 *Gallium is an unusual metal. It has a low melting point and melts when heated by your hand.*

If a material is a good conductor of heat then heat will pass through it quickly. Bad conductors of heat are called insulators.

Activity 5.7: Testing conduction of heat

Grace and Rio carried out an investigation to find out if metals and non-metals are good or bad conductors of heat.

This is the method they used:

A Use rods made out of different materials: some metals and some non-metals.

B Stick drawing pins along one rod using a small bit of candle wax.

C Heat one end of the rod. Time how long it takes for each drawing pin to fall off. Record your results.

D Repeat this with the other rods. Make sure you stick the drawing pins at the same distances along the rods.

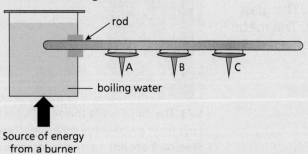

5.26 *Equipment used to find out how good different materials are at conducting heat.*

Here are their results:

Rod material	Time taken for drawing pin to fall off (seconds)		
	A	B	C
Plastic	15	600+	600+
Copper	4	7	9
Wood	32	600+	600+
Steel	8	13	19

A1 Other than the distance between the pins, name *three* control variables the students used.

A2 Why did the drawing pins fall off?

A3 Draw conclusions. For each conclusion you write, describe how the results provide evidence for this.

    a) Which material was the best conductor of heat?

    b) Are metals or non-metals the best conductors of heat?

    c) Are some metals better conductors of heat than others?

Testing conduction of electricity

To test if a material is a conductor of electricity, you can place it in an electrical circuit with a battery.

If the material conducts electricity the circuit is completed and the bulb will light up.

5.27 *The circuit contains a gap. The light bulb is off.*

Activity 5.8: Testing conduction of electricity

Grace and Rio then carried out an investigation using the equipment shown in figure 5.28.

Their results are shown in the table below.

Material	Type of material	Was the bulb on or off?
Wooden stick	Non-metal	off
Iron nail	Metal	on
Plastic straw	Non-metal	off
Pencil with wires connected to the grey graphite inside	Non-metal	on
Copper strip	Metal	on

Table 5.2 *Results from the experiment.*

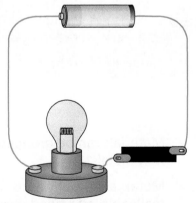

5.28 *The light bulb is on. The material can conduct electricity.*

Use the results table to answer these questions:

A1 Which of these materials conducts electricity?

A2 Rio concluded that some non-metals can conduct electricity. Describe the evidence from the results that supports his conclusion.

A3 Grace concluded that all metals conduct electricity.

a) Explain why she cannot be sure from this evidence.

b) Suggest what Grace could do to be more confident in this conclusion.

Using evidence to make conclusions

Results from investigations are evidence that all metals are good conductors of heat and electricity.

Non-metals are usually bad conductors of heat and electricity. This means heat and electricity do not pass through them easily. They are insulators.

1 Draw a table to compare the properties of metals and non-metals.

2 Use the evidence to say if each material listed below, A–C, is a metal or a non-metal.

For each, give reasons for your answer.

A: A hard and shiny solid. Good conductor of heat and electricity.

B: A shiny liquid. When put into a gap in an electrical circuit, it makes a connected loop.

C: A shiny solid. Used as an electrical insulator.

3 The cables coming out of a television are made of copper covered in plastic.

Suggest why these two materials are chosen.

Key facts:

✔ Materials can be grouped as metals or non-metals.

✔ Metals are usually hard, strong, malleable, and good conductors of heat and electricity.

Check your skills progress:

I can decide if a material is a metal or non-metal based on its physical properties.

I can make conclusions from evidence.

End of chapter review

Quick questions

1. Materials in which state of matter can be compressed? **[1]**

   a only liquids   **b** only gases

   c liquids and gases   **d** solids, liquids and gases

2. Describe the difference in movement of particles in solids and liquids. **[2]**

3. Name *two* changes in state where the energy of the particles increases. **[2]**

4. At what temperature will a liquid freeze? **[1]**

   a at 0 °C   **b** at its melting point

   c at its boiling point   **d** at any temperature below 100 °C

5. Which one of these is not a physical property of materials? **[1]**

   a hardness   **b** colour

   c how quickly it reacts with water   **d** how well it conducts electricity

6. A plastic called polythene is used to make shopping bags.

   Name *two* physical properties that make polythene suitable for this use. **[2]**

7. Some materials can conduct heat. Describe what this means. **[1]**

8. Magnesium is a metal. Suggest *three* physical properties of magnesium. **[3]**

Connect your understanding

9. If you cool a gas enough it will turn into a liquid.

   Explain what happens to the particles in the gas during this process. **[2]**

10. Copper is a malleable metal that is a good conductor of heat.

    Explain how these physical properties make it useful as a material to make cooking pans. **[2]**

11. Two students were discussing what happens to a solid when it melts.

    Jack: The solid particles will turn into liquid particles.

    Mira: The particles will stay the same.

    (a) Which student is correct? **[1]**

    (b) Describe what happens to the particles when a solid melts. **[3]**

12. Amir heated a solid, X. He measured the temperature every 30 seconds.

His results are shown in the table.

Time (s)	0	30	60	90	120	150	180	210	240
Temperature of X (°C)	22	28	45	59	59	59	66	79	84

(a) Estimate the melting point of X. Explain how you used the results to decide this. [2]

(b) Amir concludes that X is probably a non-metal. Suggest why he thinks this, based on the evidence he has collected. [1]

(c) Describe another experiment he can carry out to collect evidence to support his conclusion. [4]

13. A cup of water was left in a room overnight.

Predict what would happen to the volume of water in the cup.

Give a scientific explanation for your prediction. [4]

Challenge question

14. The melting point of iron is much higher than the melting point of water.

Suggest why. Use particle theory in your answer. [3]

6

Chapter 6
The Earth

What's it all about?

Our Earth formed over 4.5 billion years ago. Most of the Earth is made from rock, but it has a large metal ball at its centre.

Scientists study rocks to find out what the Earth is made from, how it has changed and how life on Earth has changed. Some rocks contain the remains of plants and animals from millions of years ago. At these cliffs in Canada the world's oldest known fossil reptile was found along with fossils of extinct giant trees, ferns and insects from 300 million years ago.

You will learn about:
- How different types of rocks form
- What fossils are and what they tell us about how long life has existed on Earth
- A simple model of the structure of the Earth
- How scientists can estimate the age of the Earth

You will build your skills in:
- Observing and classifying materials
- Choosing the right variables to change, control and measure in an experiment
- Presenting conclusions from collected data
- Researching current scientific thinking using secondary sources

Rocks and soils

Learning outcomes

- To observe and classify different types of rocks and soils
- To describe how different rocks form
- To choose variables to change, control and measure in an experiment

Starting point

You should know that...	You should be able to...
There are different types of rock	Present data using tables
Everyday materials, including rocks and soil, can be sorted into groups because of their properties	Plan how to carry out a fair test
Scientists combine evidence from observation and measurements to suggest new ideas	Use results to make conclusions

Rocks

6.1 *Not all rocks look the same.*

Rocks are very important. We use them all the time, to build houses, buildings and statues.

Rocks are made from different **minerals**. These minerals have different shapes and colours and are made of different substances, which is why rocks look so different and have different properties.

Some rocks have layers and some have **crystals**.

 1 Make a list of *five* different ways that rocks can be used.

Key terms

crystal: solid in which particles are arranged in a regular pattern.

mineral: solid substance with a fixed chemical composition. Most minerals are crystals.

Activity 6.1: Observation skills: what can you see?

Your teacher may give you some rock samples to look at carefully. Alternatively, look carefully at the following rock photographs.

A1 Write *at least one* observation for each rock.

Use the following key terms to help you: *large crystals, small crystals, layers, light, dark, rough, smooth, colours, pattern, fossil.*

A2 Use the similarities and differences that you observe to sort the rocks into groups. Be ready to talk about how you decided to put the rocks into different groups.

6.2 *Granite*

6.3 *Sandstone*

6.4 *Marble*

6.5 *Conglomerate*

6.6 *Slate*

6.7 *Basalt*

6.8 *Limestone*

Properties of rocks

Rocks also have different properties. For example, the **porosity** of the rock, how it reacts with acid or how hard it is.

You can use these differences in properties to classify rocks into groups.

Key term

porosity: the amount of empty space in a material.

Activity 6.2: Testing the properties of rocks

You are going to investigate the properties of some different rocks by testing them. For example, the **scratch test** is done to test how hard a rock is.

Your teacher will show you how to carry out the tests before you start.

You must draw a results table to record your results.

Remember: write down the name of the rock and describe what it looks like.

A1 Which rock was the hardest? How did you reach this conclusion?

A2 Which rock was the most porous? How did you reach this conclusion?

2 Which rock is most likely to contain oil or gas? Why?

Igneous rocks

Rocks are classified by how they formed.

Igneous rocks are sometimes called volcanic rocks. They form when molten rock called **magma** or **lava** cools and solidifies.

Igneous rocks can form above or beneath the Earth's surface. They contain crystals. There are no spaces between the crystals, which means that igneous rocks have a low porosity.

The size of the crystals depends on how slowly the magma cools and solidifies. If the magma cools underground, it cools and solidifies very slowly. This can take thousands of years. This means the crystals have time to grow, so they are large.

Granite is an igneous rock formed in this way.

If the magma is erupted at the surface of the Earth, for example at a volcano, it is called lava. Lava at the Earth's surface cools and solidifies quickly. This means the crystals do not have time to grow and the crystals are small.

Basalt is an igneous rock formed in this way.

Key terms

igneous rock: rock formed when magma cools and solidifies.

lava: magma at the Earth's surface.

magma: molten rock found below the Earth's surface.

scratch test: test to see how hard a rock is, by how easy it is to scratch.

6.10 *A volcano erupting.*

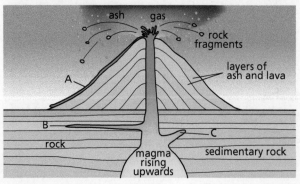

6.9 *Cross section of a volcano.*

3 Look at figure 6.9. Would you expect rocks with small crystals to form at A, B or C?

4 Would you expect rocks with large crystals to form at A, B or C?

5 Explain why granite has large crystals and basalt has small crystals.

Activity 6.3: How do igneous rocks form?

You are a particle in an igneous rock.

Describe how your igneous rock formed and include in your description:

- where you formed
- what you formed from – lava or magma
- how big your crystals are
- how long it took to cool down.

Your description could be a story, or a series of pictures with explanations.

Danger! Volcano erupting

Krakatoa erupted in August 1883 and was one of the most deadly volcanic eruptions of modern history, killing at least 35 000 people.

There were four major eruptions. The fourth made the loudest sound ever recorded on the planet. It was heard from 3000 miles away.

Most of the deaths that occurred when this volcano erupted were actually due to enormous waves (tsunamis) which destroyed coastal villages and towns.

6.11 *Krakatoa, a volcano in Indonesia that is still active.*

Look at figure 6.12. This is another type of igneous rock called pumice, which is sometimes made when a volcano erupts. Pumice floats on water.

6 What properties does pumice have, that make it different from other igneous rocks?

7 Why do you think the rock has holes in it?

6.12 *Pumice, a volcanic igneous rock.*

Sedimentary rocks

Sedimentary rocks are another type of rock. They form in layers, as you can see in figure 6.13. Sedimentary rocks are formed when other rocks are broken down into smaller pieces and are transported by water, wind or ice. When the pieces of rock stop moving, they are deposited (dropped) as **sediments.**

6.13 *Layers in a sedimentary rock.*

More and more sediments are dropped and layers start to form. The weight of the sediments compresses (squashes) the layers below so any water is squeezed out.

Chemicals form that hold the sediment particles together and the layers gradually turn into sedimentary rock.

Sandstone is a type of sedimentary rock which is made from sand grains.

Sedimentary rocks often contain the traces or remains of dead organisms. These remains are called **fossils**.

Limestone is a type of sedimentary rock that often contains the shells and skeletons of dead organisms that lived in the sea.

6.14 *Dead plants and animals trapped in sediments may change into fossils over millions of years.*

6.15 *Uluru (Ayers Rock), a huge sandstone rock in the middle of Australia.*

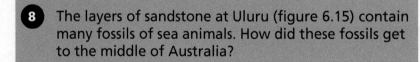

8   The layers of sandstone at Uluru (figure 6.15) contain many fossils of sea animals. How did these fossils get to the middle of Australia?

Key terms

fossils: the traces or remains of dead organisms that lived thousands or millions of years ago.

sedimentary rocks: rocks formed from layers of sediment deposited by water, wind or ice.

sediments: small pieces of rock, such as pebbles, sand and mud.

Metamorphic rocks

Metamorphic rocks are made when sedimentary or igneous rocks are changed over many thousands of years by very high temperatures or very high pressure (squeezing) or both.

Metamorphic rocks are made of crystals. The rocks are hard and the crystals have no spaces between them, so they have a low porosity.

6.16 *Marble, a metamorphic rock which is often polished and used on floors and to make tiles.*

The sedimentary rock, limestone, changes into marble when it is heated to high temperatures. The limestone does not melt, it forms solid crystals.

9 Could marble contain fossils? Why?

Figure 6.17 shows a sample of slate. This is another example of a metamorphic rock. It is formed when mudstone, a sedimentary rock made from clay, is compressed over thousands of years.

10 Describe how igneous rocks are changed into metamorphic rocks.

Soil

Soil is one of the Earth's most important resources. It is needed for growing plants to feed animals and people.

Soil is a mixture of different substances formed in layers. It is made of small particles of rock, dead animals and plants, water and air.

The top soil is made mostly from sand and clay particles. Its rich brown colour comes from decaying twigs and leaves from plants. This is called **humus**.

The next layer contains mostly pieces of rock.

The three main types of soil are clay, loam and sandy soil.

- Clay soil is made up of very small rock particles which fit together closely, so there are not many spaces.
- Sandy soil has bigger rock particles with bigger spaces between them.
- Loam soil is a mixture of clay and sand.

Key term

metamorphic rock: rock formed when sedimentary or igneous rocks are changed by very high temperatures and/or pressure.

6.17 *Slate, a metamorphic rock.*

6.18 *The layers in soil.*

Key terms

humus: the part of soil which is made from dead or rotting plant material.

soil: mixture of small particles of rock, dead animals and plants, water and air.

Activity 6.4: Investigating soil content

A scientist wants to find out how much water is in different soils. They weigh soil samples on a balance and record their masses.

The soil sample is then put into a warm oven for one day.

Water evaporates from the soil when it is in the oven.

The soil is weighed again and the mass recorded.

The results of the experiment are shown in the table below.

Soil sample	Mass at start (g)	Mass after warming (g)	Change in mass (g)
A	100	95	5
B	101	96	
C	100	94	
D	100	94	
E	102	92	

A1 Complete the table. One row has been done for you.

A2 Draw a bar chart to show your results.

A3 Which soil sample contained the most water?

A4 Why do you think the amount of water was different for different soil samples?

A5 Why is the water content of a sandy soil usually much less than for a clay soil?

A6 How could the scientist make the data from the investigation more **reliable**?

Activity 6.5: Investigating soil draining

How quickly does water pass through soil?

Plan an investigation to find out how quickly water will pass though different types of soil.

In your investigation, you will change one **variable**, measure another variable and control all other variables.

When producing your plan you should include the following:

- a list of the apparatus you will use
- a list of the things you will do – your method
- what you will change (the **independent variable**) and what you will measure (the **dependent variable**)
- a list of the things you will keep the same (the control variables)
- a table to record your results.

A1 Carry out your investigation and record your results.

A2 Draw a bar chart to show your results.

A3 Which type of soil drained fastest?

A4 Why do you think it drained fastest?

Key terms

dependent variable: variable you decide to measure in an experiment.

independent variable: variable you decide to change in an experiment.

reliable: measurements are reliable when repeated measurements give results that are very similar.

variable: something that may change.

Growing rice

The world's three biggest food crops are rice, wheat and maize.

Rice is a main food for over half the world's population. Rice plants need lots of water to grow well. Clay soils are best to grow rice in because these soils stay wet for longer. The water does not drain away.

6.19 *Rice growing in a field with very wet soil.*

Key facts:

✔ Rocks are put into three groups – igneous, sedimentary and metamorphic.

✔ Igneous rocks are formed when magma (molten rock) cools and solidifies.

✔ Sedimentary rocks are formed from layers of sediment.

✔ Metamorphic rocks are formed when existing rocks are changed by very high temperatures and/or pressure.

✔ Soil is a mixture of particles of rock, dead animals and plants, water and air.

✔ There are different types of soil which hold different amounts of water.

Check your skills progress:

I can use observations to put rocks into different groups.

I can plan investigations and think about which variables to change, observe and control.

I can present results in tables and bar charts.

I can use evidence to support my conclusions.

Fossils and the fossil record

- To describe how fossils form
- To research what the fossil record is and what it can tell us
- To explain how the fossil record can tell us about the age of the Earth
- To learn about the most recent estimates of the age of the Earth

Starting point

You should know that...	You should be able to...
Scientists use their observations to suggest new ideas	Use different sources to research information
Scientists collect evidence to test these ideas	

What do fossils tell us?

Palaeontologists are scientists who study fossils so they can learn about plants and animals that lived at different times. They also study how plants and animals have changed over long periods of time.

1. Fossils are often found in sedimentary rocks. Why are they not found in igneous rocks?

2. Different layers of sedimentary rock have different fossils. In which layer would you expect to find the oldest fossils?

Most fossils are the preserved remains of dead organisms, but sometimes the tracks made by an animal can be preserved. The dinosaur footprints in figure 6.21 were made in soft, wet sediment. As the sediment dried the tracks hardened. The tracks were then buried by more sediment, which eventually hardened into rock. As the rock has been worn away over millions of years, the footprints have become visible.

Tooth marks and burrows can be preserved in the same way. These fossils give information about how the animal lived and behaved.

6.20 *A rock containing trilobite fossils – these are found on every continent on Earth. Trilobites became extinct 250 million years ago.*

Key term

palaeontologist: scientist who studies fossils.

6.21 *Fossilised dinosaur tracks in Kalasin, Thailand.*

The fossil record

The collection of fossils from different times and places in the Earth's past is called the **fossil record**. These fossils provide **evidence** of how different groups of animals and plants have changed over millions of years. Most fossils are the remains of organisms which are now **extinct**.

Millions of years before present

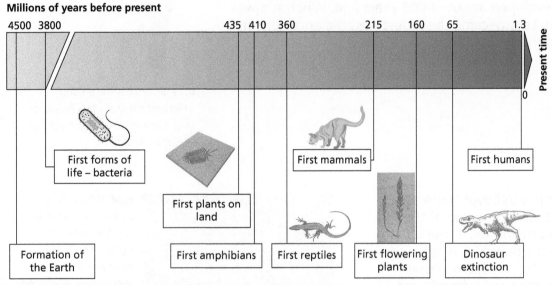

6.22 *A timeline showing when different kinds of organisms first appeared on Earth.*

There are sometimes gaps in the fossil record where scientists do not know what the animal or plant was like at that time, because there are no fossils. These gaps can happen if the remains of the organism are not preserved or if the rocks containing fossils are deeply buried or have been worn away. If an animal has a soft body (no skeleton), its remains are unlikely to be preserved as a fossil.

3 Squid are soft-bodied sea creatures. Why would squid fossils be difficult to find?

4 Name some modern sea animals that could form fossils.

5 Explain how a fossil forms. Use these words – buried, pressure, sea, sedimentary rock.

6 Why are there gaps in the fossil record?

Key terms

evidence: data or observations we use to support or oppose an idea.

extinct (life forms): does not exist any more.

fossil record: collection of fossils identified from different times in the Earth's past that shows how different groups of animals and plants have changed over millions of years.

Who was Lucy?

Fossils are found in all parts of our world, including the fossil bones of our human ancestors.

Lucy is one of the most famous human-like fossils. She was found in Ethiopia in 1974 and scientists believe that she is over 3 million years old.

Until the 1950s, scientists believed the ancestors to modern humans developed about 60 000 years ago. When she was discovered, Lucy became the oldest possible ancestor for modern humans.

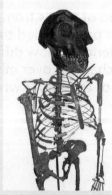

6.23 *Scientists have used information from the fossils of Lucy's bones to reconstruct her skeleton and skull.*

At Whale Valley in Egypt there are hundreds of fossils of some of the earliest types of whales. These whales had tiny back legs, which whales now do not have. This is evidence that whales have changed over millions of years, and that the ancestors of whales walked on land.

6.24 *Whale fossil at Wadi El Hitan or 'Whale Valley' in Egypt. The fossils here are found in sandstone, which was once at the bottom of the sea.*

Activity 6.6: Evidence from the fossil record

Research another example where the fossil record shows changes in a group of animals, like the whale fossils found in Egypt, or the extinction of a species.

Remember that all scientific books and papers are always checked to make sure that what has been written is correct. Information on websites is not usually checked like this. So, be careful about the sources you use.

Think about:

- Are the fossils related to a modern animal?
- What are the oldest fossils like?
- What are the youngest fossils like?
- How does this prove that the species has changed over time?

A1 Make sure you choose your sources carefully – ask your teacher for some websites or books to get you started if you are stuck.

A2 Present your research as a short report (about 200 words) or digital presentation.

A3 Include suitable images.

A4 Record your sources of information.

Age of rocks and age of the Earth

Sedimentary rocks are built up in layers. Usually, the deeper layers contain older fossils because the deeper layers were deposited first. This helps to show us how a species changed over time.

By comparing the fossils found in rocks in different places scientists can compare the age of the rocks. This is because layers of rock containing the same fossil species must be the same age.

Scientists can find the actual age of a rock and any fossils in that rock using chemical tests. This is how we know, for example, that the oldest plant fossils are older than the oldest animal fossils. The method gives a small range of estimated age for a rock, such as 100 million years ± 5 million years.

6.25 *A plant fossil.*

> **7** How can fossils help scientists to find out how old, in years, a rock is?

Scientists use the results from their own experiments as evidence when they make conclusions about the age of rocks and fossils. They also use evidence from **secondary sources** like books and scientific papers. These sources include results from experiments that other scientists have done.

Scientists know that the Earth is approximately 4.5 billion years old, because many different chemical tests have been done on different rocks around the world. This evidence has been checked by many scientists.

> **8** The world's oldest known fossils are the remains of bacteria that lived about 3.8 billion years ago. Does the age of these fossils tell you the age of the Earth?

> **9** Why can scientists not use the internet alone for finding out about the age of the Earth?

Key term

secondary sources: information that has been produced by somebody else.

Key facts:

✔ Sedimentary rocks sometimes contain fossils – igneous rocks do not.

✔ Fossils are the remains of animals and plants from millions of years ago.

✔ The fossil record provides evidence of how living things changed gradually over time.

✔ The Earth is more than 4 billion years old.

✔ The oldest fossils are almost as old as the Earth itself, showing that life started billions of years ago.

Check your skills progress:

I can choose appropriate sources to research information.

I can identify evidence in these sources.

Structure of the Earth

Learning outcomes
- To use models to describe the internal structure of the Earth

Starting point

You should know that...	You should be able to...
Scientists use their observations to suggest new ideas	Choose appropriate sources to research scientific ideas
Scientists collect evidence to test these ideas	

Structure of the Earth

It is not possible to travel to the centre of the Earth, so it is difficult to know what is below the Earth's surface. It is not possible to drill deeper than 12 km – that is 0.2% of the Earth's depth.

By studying earthquakes, scientists can make a model about what is inside the Earth. The data from earthquakes provides evidence that there are different layers. We can use this data to create a model of how the Earth is structured. Models are very useful for scientists, because they help us to understand something which we cannot see.

The Earth is made up of four different layers – the **inner core**, **outer core**, **mantle** and **crust**.

The crust and mantle are rock, but the core is metal – it is made from nickel and iron.

Figure 6.26 shows a model of the structure of the Earth. The crust varies in thickness from around 7 km to 70 km deep and the radius of the whole Earth is about 6400 km.

Key terms

crust: the thin outer layer of the Earth.

inner core: solid layer of the Earth, made of nickel and iron.

mantle: the layer of the Earth beneath the crust. It is mostly solid but it can flow very slowly.

outer core: liquid layer around the inner core of the Earth, made of nickel and iron.

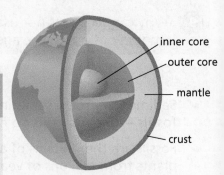

6.26 *Model of the structure of the Earth, showing the four layers.*

Activity 6.7: Researching the Earth's internal structure

What are the layers that make up Earth like?

Use books and the internet to research what scientists think the different layers are like. Create a poster to describe the layers that make up the Earth. Include a diagram.

Think about:

A1 What is the main type of rock found in the Earth's crust?

A2 Are the other layers solid or liquid?

A3 How hot is each layer?

Evidence from meteorites

Meteorites are fragments from asteroids that fall onto the Earth's surface from space. The Earth and other planets in the Solar System formed at the same time as asteroids, from the same material. So scientists use what meteorites are made of to model what the interior of the Earth is made of. Some meteorites are made from rock but some are nearly all iron.

Meteorites that originate from asteroids are all about 4.5 billion years old.

6.27 *A close up of part of the 60-ton Hoba meteorite from Namibia. This is the largest piece of iron ever found near the Earth's surface.*

 1 Suggest a reason why iron and nickel make up the core of the Earth.

A jigsaw

The Earth's outer layer is split into about 20 large pieces, called plates.

The plates are moving very slowly all the time. They sometimes cause volcanoes and earthquakes when they push against each other or move over or under each other. The Himalaya Mountains are getting 1 cm taller every year because two plates are pushing against each other.

6.28 *The tectonic plates.*

As the tectonic plates move in different ways, some rocks at the surface are pushed deep into the crust and then changed (metamorphosed) into different rocks. Some rocks are even pushed deeper, into the mantle, and melted. Scientists say that the rocks that were at the surface are recycled. This is why most rocks on the Earth's surface are much younger than the Earth itself.

Key facts:

✔ There are no rocks at the surface from deep inside the Earth, so scientists have to use data from earthquakes to make a model of its structure.

✔ The Earth has four layers – the inner core, outer core, mantle and crust.

Check your skills progress:

I can choose sources of information to answer a research question.

End of chapter review

Quick questions

1. All rocks can be classified into one of three main groups. State what these are. **[3]**

2. Each of the three groups of rock is formed in a different way. For rocks A–C, state the type of rock: **[3]**

   (a) Rock A: formed in the sea and contains fossils

   (b) Rock B: formed when another type of rock is changed due to very high temperature and pressure

   (c) Rock C: formed when hot magma cools and solidifies

3. Rocks all look different. Put ticks (✓) in the following table, to show which key features may be present in each of the different rock types. **[3]**

Feature	Igneous	Sedimentary	Metamorphic
Crystals			
Layers			
Fossils			

4. State which letter, A–D, in figure 6.29 represents the following:

   (a) Lava =

   (b) Cooled and solidified rock =

   (c) Magma =

   (d) Ash = **[4]**

5. What is the difference between magma and lava? **[2]**

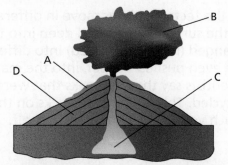

6.29 *Diagram of a volcano.*

6. The list below gives four steps in the formation of sedimentary rock.
   Put the steps in the correct order. **[2]**

   A The sediments are covered by more sediment.

   B The layers of sediments become stuck together and harder.

   C The upper layers press on the lower layers.

   D Small pieces of sediment produced from larger pieces of rock are deposited.

7. What is a fossil? [1]

8. Name the four layers of the Earth. [4]

9. Complete the table below to show which layer of the Earth is being described. The first one has been done for you. [3]

Letter	Description	Layer
A	Made from semi-molten rock	Mantle
B	Made from solid rock	
C	Made from solid iron and nickel	
D	Made from molten iron and nickel	

Connect your understanding

10. Explain why igneous rocks that form deep underground have different sized crystals from igneous rocks that form at the surface. [3]

11. Where are the oldest layers of rock usually found in a cliff made from sedimentary rocks? Why is this the case? [2]

12. There are different types of soil which contain different amounts of clay or sand.

(a) Compare how well clay soil and sandy soil hold water (stay wet). [1]

(b) Explain your answer. [3]

13. Different types of rock form in different ways.

(a) State how igneous, sedimentary and metamorphic rocks are formed. [3]

(b) State which type never contains fossils and explain why. [2]

14. Explain why there are gaps in the fossil record. [3]

Chapter 7

Acids and alkalis

What's it all about?

Acids and alkalis are used in our lives every day – you might not even realise when you are using them. They are found in food such as oranges and lemons, cleaning products and medicines.

Acids and alkalis are also very important in your body. Some cells in the body produce acids and alkalis to help the body function. The body also excretes substances which are too acidic or alkaline to keep the balance in the body right.

You will learn about:
- How to tell if a solution is an acid or an alkali
- Why acids and alkalis are useful

You will build your skills in:
- Suggesting ideas that may be tested
- Planning investigations and considering what variables to change or keep the same
- Choosing the right equipment to investigate a question
- Making conclusions from your results

Acids, alkalis and the pH scale

Learning outcomes

- To use an indicator to find out whether something is an acid or an alkali
- To use the pH scale

Starting point

You should know that...	You should be able to...
Some materials can dissolve in water	Talk about risks and how to avoid danger
	Draw tables to show your results

Acids and alkalis

Acids and **alkalis** can be found in many everyday substances. Acids and alkalis can be useful, for example in cleaning products. But they can also be harmful.

Acids and alkalis found in fruits, milk and toothpaste are weak and are harmless. However, some acids and alkalis are strong acids or strong alkalis. This means that they can cause harm and you must be very careful when handling them. A common acid used in laboratories is hydrochloric acid and this acid is much stronger than any acid you would find in food. A common alkali used in laboratories is sodium hydroxide and this alkali is stronger than any alkali you would find in soap.

If a substance causes harm, **hazard symbols** are used to warn people so they can take precautions such as wearing eye protection and handling the substance safely.

Strong acids or alkalis are **corrosive** – they burn the skin. Acids or alkalis that are not corrosive can still be harmful if they irritate the skin (make it red) or the eyes.

Key terms

acid: substance which has a pH of less than 7 on the pH scale.

alkali: base that dissolves in water to make a solution with a pH of more than 7. See page 150 for definition of base.

corrosive: substance that causes burns to skin and eyes and damages other materials.

hazard symbol: symbol which warns you about the dangers of an object, substance or radiation.

7.1 Moderate hazard symbol (for example, causes skin or eye irritation, or harmful if swallowed or inhaled).

7.2 Corrosive hazard symbol.

1. Why is it important for some chemicals to have hazard symbols on them?

2. Why do scientists use symbols instead of writing words?

3. A chemical could cause burns to the skin and eyes if not used safely. What hazard symbol would you put on this chemical?

4. A chemical could cause the skin to become red and itchy if it is in contact with the skin. What hazard symbol would you use for this chemical?

Activity 7.1: Hazard symbols

Find labels on some chemicals that you or your family use at home. These could be cleaning products, soaps or shampoo, food or fizzy drinks, DIY products or any others you can think of. You should present your findings in a table.

A1 What hazard symbols, if any, does the label have? What does it mean if there is no hazard symbol?

A2 Does it have any acids or alkalis in it? (Clue – most acids have the word 'acid' in their name. A lot of alkalis have 'soda' or 'hydroxide' in their name. Ammonia solution is also an **alkaline** solution.)

A3 What harmful effects, if any, do the substances you have chosen have?

7.3 *Hazard warning on a bottle of bleach.*

Indicators

How do you know whether a substance is an acid or an alkali?

Scientists use chemicals called **indicators** to tell whether something is an acid or an alkali. Indicators change colour and this shows you whether a substance is **acidic** or alkaline.

Litmus is an indicator solution which is red in an acid and blue in an alkali. Litmus paper is paper soaked in litmus and then dried. You dip the paper into a sample of the solution and it changes colour, telling you whether the solution is acidic or alkaline. Blue litmus paper turns red in an acid but stays blue in an alkali. Red litmus paper turns blue in an alkali but stays red in an acid.

Key terms

acidic: having the properties of an acid.

alkaline: if a base is dissolved in water then the solution is alkaline.

indicator: chemical that changes colour in an acid or alkali.

litmus: type of indicator which turns red in an acid and blue in an alkali.

5 Copy and complete the following table on litmus indicator colours in acids and alkalis.

	Colour of red litmus paper	Colour of blue litmus paper
In acidic solution		
In alkaline solution		

6 What colour would blue litmus paper turn in hydrochloric acid?

7 What colour would red litmus paper turn in sodium hydroxide?

7.4 *Blue and red litmus paper in acidic solution (left). Blue and red litmus paper in alkaline solution (right).*

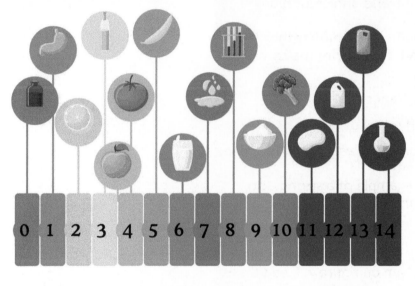

0 1 2 3 4 5 6 7 8 9 10 11 12 13 14

acidic   neutral   alkaline

7.5 *A colour chart for universal indicator using the pH scale.*

7.6 *Using universal indicator to measure the pH of a solution.*

Universal indicator is an indicator that can change into many different colours, from red through yellow and green to purple. The colour change of universal indicator is used to measure how strong or weak the acid or alkali is on the **pH scale.**

Different substances have different pH values depending on how strong or weak an acid or alkali they are or if they are **neutral**.

Scientists look at the colour change of the universal indicator and match it to the pH scale. If the colour is green then the pH is 7, which means the solution is neutral. The closer to a pH of 0 a solution is, the more acidic it is. The closer to a pH of 14 a solution is, the more alkaline it is.

Key terms

neutral: neither acid nor alkali. If soluble, it produces a solution of pH 7.

pH scale: scale from 0 to 14 which measures how strong or weak an acid or alkali is.

universal indicator: type of indicator which can change into a range of colours depending on whether the solution is acidic or alkaline and how strong it is.

8 Acids and alkalis can have a range of pH values depending on whether they are strong or weak.

   a) What range of pH does a strong acid have?

   b) What range of pH does a weak alkali have?

9 State whether the following are acids, alkalis or neutral substances. If they are an acid or an alkali, you should also say whether they are strong or weak.

   a) A substance with a pH of 3.

   b) A liquid with a pH of 8.

   c) A substance which turns universal indicator yellow.

   d) A chemical with a pH of 7.

   e) A substance with a pH of 14.

   f) A liquid which turns universal indicator blue.

10 For the following statements, say whether they are true or false. If they are false, write the correct answer.

   a) A strong acid has a pH range of 4–6.

   b) A weak alkali has a pH range of 8–10.

   c) A neutral substance turns universal indicator green.

11 Using figure 7.5, give an example of a strong acid, weak acid, strong alkali and a weak alkali.

12 Draw and colour in your own pH scale and label each pH value with an example. If you can find some other examples which are not given on figure 7.5, add these.

13 Using books or the internet, find the names of at least *two* other indicators that scientists use.

14 Why is universal indicator more useful than other indicators?

15 What is a disadvantage of using litmus indicator?

Activity 7.2: Planning an investigation – is it an acid or an alkali?

You are going to produce a plan to find out whether some household substances are acids, alkalis or neutral and what their pH is. You might think you know what the pHs of some of the substances are already. Your plan should show how you will find out the pH values and prove whether your predictions are correct.

The substances you must test are:

- shampoo
- vinegar
- orange juice
- toothpaste
- tap water
- indigestion tables (dissolved in water).

A1 Make a prediction for each of the substances – is it an acid, alkali or neutral? What pH do you think it will have?

A2 Write a list of equipment you will need to use for this experiment. Include any safety precautions.

A3 Write a method you would use – think carefully about which indicator you would use and explain why you have chosen this indicator.

A4 Draw a results table you would use for this investigation.

Acid rain damage

When the ancient stone temples were built in Cambodia, acid rain was not a problem. The temples are now becoming damaged due to acid rain.

Rain can become acidic when the gases released when fossil fuels are burned dissolve in water vapour in the atmosphere. Burning more coal in power stations has made the problem of acid rain much worse in the last 40 years.

Acid rain occurs everywhere, not just in Cambodia. Buildings across the world have been damaged by acid rain. Acid rain also damages statues, damages forests and can kill animals and fish.

7.7 Ancient stone temples in Cambodia.

Key facts:

✔ Acids and alkalis can be found in many everyday substances.

✔ The pH scale goes from 0 to 14 and shows how acidic or alkaline a solution is.

✔ The closer to 0 the pH is, the more acidic that solution is.

✔ The closer to 14 the pH is, the more alkaline the solution is.

✔ Neutral solutions have a pH of 7.

✔ An indicator changes colour in a substance if it is acidic, alkaline or neutral.

Check your skills progress:

I can plan investigations and choose appropriate equipment.

I can draw results tables for an investigation.

I can use evidence to support my conclusions.

Neutralisation

- To use indicators and the pH scale to find out if a solution is acidic, alkaline or neutral
- To understand what neutralisation means
- To give examples of when neutralisation is useful

Starting point

You should know that...	You should be able to...
Some materials can dissolve in water	Measure the volume of a liquid
When a solid dissolves in water a solution is formed	Use results tables
	Design a fair test

What is neutralisation?

If you add an acid to an alkali a chemical reaction happens. Water is formed, as well as another substance which may dissolve in the water.

If you mix exactly the right amount of acid and alkali a neutral solution is formed. This type of reaction is called **neutralisation**.

A **base** is any substance that neutralises an acid. If a base dissolves in water, then it is called an alkali.

The pH changes during a neutralisation reaction. An acidic solution becomes less acidic when you add a base or alkali by a small amount at a time. The pH increases towards 7. When the pH reaches 7 you have a neutral solution. If you add more base or alkali to the acid, the solution becomes more alkaline and the pH increases beyond 7.

Key terms

base: substance that neutralises an acid. It has a pH of more than 7 on the pH scale.

neutralisation: chemical reaction between an acid and a base which produces a neutral solution.

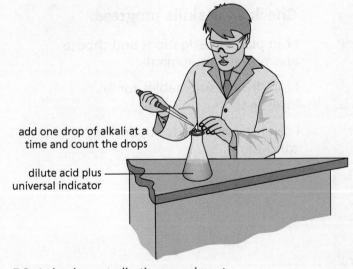

add one drop of alkali at a time and count the drops

dilute acid plus universal indicator

7.8 *A simple neutralisation experiment.*

1. What pH does a neutral substance have?

2. Sometimes, acidic or alkaline solutions need to be neutralised. Describe how you could neutralise an acidic solution and how you would know that neutralisation had happened.

3. What colour would universal indicator change to if added to a neutral solution?

4. The simple experiment shown in figure 7.8 could be changed by using a measuring cylinder to add the alkali, and using the colour change of the indicator to estimate pH. How could you show the way that the pH changes as you neutralise the acid with an alkali?

Uses of neutralisation

Neutralisation can be a very useful chemical reaction in everyday life. One example is to treat indigestion.

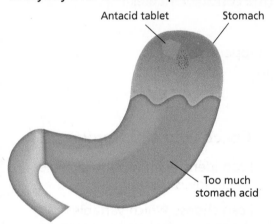

Antacid tablet   Stomach

Too much
stomach acid

7.9 *An antacid tablet contains a base, which will neutralise the extra acid in the stomach.*

Your stomach produces a strong acid to help digest the food that you eat and to protect against infection. Sometimes there can be too much of this stomach acid or it can move up from your stomach into your throat. When this happens it can be very painful. People may take a substance called an antacid or indigestion tablet to stop the pain.

Antacids or indigestion tablets contain a base to neutralise the stomach acid. Neutralising the acid gets rid of the pain.

Another example of neutralisation is when farmers add lime to acidic soil. This makes the soil less acidic so that their crops can grow better.

7.10 *A farmer spreading lime on a field to make the soil less acidic.*

Activity 7.3: Neutralisation

A1 Use different books and the internet to research other examples of neutralisation being useful in everyday life. You should show your findings using a short presentation (5 minutes maximum). Make sure that you record all sources of information that you use.

Activity 7.4: Investigating neutralisation

Plan an investigation to find out which of three acids is the strongest. You will use a solution of an indigestion tablet dissolved in water to neutralise the acids.

In your investigation, you will change some variables, stop others from changing and measure others.

A1 When producing your plan you should include the following:

    A A list of the apparatus you will use.

    B A list of the things you will do – your method. Include any safety precautions.

    C What you will change (the independent variable) and what you will measure (the dependent variable).

    D A list of the things you will keep the same (the control variables).

    E A table to record your results.

A2 How would you know when neutralisation has happened?

A3 How would you know which acid is the strongest?

Key facts:

✔ Neutralisation is a chemical reaction which happens when an acid and an alkali react together to make a neutral solution.

✔ Neutralisation is a very useful chemical reaction. For example, it is used to treat indigestion and neutralise acidic soil.

Check your skills progress:

I can identify important variables for an investigation.

I can choose which variables to change, control and measure.

I can choose the correct equipment to collect evidence and answer a scientific question.

End of chapter review

Quick questions

1. State the meaning of the following hazard symbols.

   (a)   **(b)**

   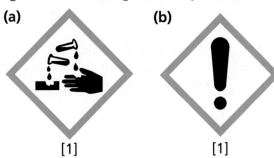

   [1]   [1]

   (c) Give the name of a substance which changes colour to show whether a
   substance is an acid, alkali or neutral.   [1]

2. What range of numbers is shown on the pH scale?   [1]

3. Copy and complete the following sentences about indicators and pH.   [2]

   Litmus is an example of an indicator. It turns different colours in acids and alkalis.
   Litmus turns _____ if it is in an acid and it turns _____ if it is in an alkali.

4. Copy and complete the following sentences about reacting acids and alkalis
   together.   [3]

   When an acid and an alkali react together they produce a _____ solution.
   This has a pH of _____. This type of reaction is called _____.

5. Universal indicator turns a range of colours in acids and alkalis.

   The table shows the pH range for the different colours of universal indicator.

Colour of indicator	Red	Orange	Yellow	Green	Blue	Purple
pH	0–3	4–5	6	7	8–10	11–14

   A student tested five substances with universal indicator solution.

   Copy the table and put *one* tick (✓) in each row to state whether the results
   show the substance is acidic, alkaline or neutral.   [5]

Substance	Colour of indicator	Acid?	Alkali?	Neutral?
Toothpaste	Blue			
Vinegar	Red			
Milk	Yellow			
Water	Green			
Oven cleaner	Purple			

6. Bees and wasps are insects which use a sting to defend themselves.

A wasp sting has a pH of 10 and a bee sting has a pH of 2.

(a) Copy and complete the following table about wasp and bee stings. [4]

7.11 *A wasp.*

	Acid or alkaline?	What colour would you see with universal indicator?
Wasp sting pH 10		
Bee sting pH 2		

(b) What would you add to a bee sting to neutralise it and stop it from hurting? Tip: Can you think of any specific examples which would be safe? [2]

(c) What would you add to a wasp sting to neutralise it and stop it from hurting? Tip: Can you think of any specific examples which would be safe? [2]

Connect your understanding

7. Explain why scientists might find universal indicator more useful than litmus. [2]

8. The table below shows the pH of four different soil samples which are labelled A–D.

soil sample	pH of soil
A	5.0
B	7.4
C	7.0
D	4.7

Use letters from the table to answer the following questions:

(a) Which soil sample is neutral? [1]

(b) Cabbage grows better in an alkaline soil. Which soil sample would be the best to grow cabbage in? [1]

(c) Potatoes grow better in slightly acidic soil. Which soil sample would be the best to grow potatoes in? [1]

9. Some students did an experiment to measure the change in pH when an alkali was added to an acid. They added an alkali to an acid a little at a time using a measuring cylinder. They measured the pH every time they added the alkali.

The students used a pH meter. This is a sensor which measures pH accurately without having to look at the colour change of an indicator.

7.12 *A pH meter.*

Volume of alkali added (cm³)	pH of reaction mixture
0	5.0
2	5.0
4	5.0
6	6.0
8	7.1
10	8.0
12	8.5
14	9.0
16	9.0
18	9.0
20	9.0

(a) Plot a line graph of the results on graph paper. Make sure that you plot the pH on the vertical (*y*) axis and volume of alkali added on the horizontal (*x*) axis. [2]

(b) Describe what you can see happening from the results on the graph. [3]

End of stage review

1. The Earth has *four* layers.

   (a) Name the layers A–D in this diagram of the Earth. [4]

   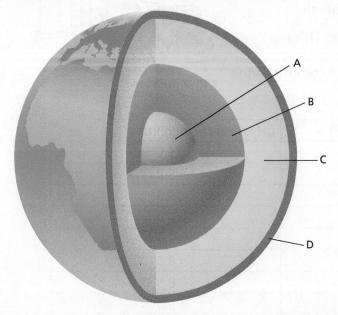

   (b) What *two* metals are found at the centre of the Earth? [2]

   (c) Which *two* sentences show the correct facts about the age of the Earth? [1]

   (i) The Earth is more than 4000 million years old

   (ii) The fossil record gives a time when scientists think the Earth formed

   (iii) The age of Earth rocks can be measured using chemical tests

   (iv) Rocks on the surface of the Earth are all the same age

   (d) There are four layers of sedimentary rock in a cliff (right).

   How can you tell that D is the oldest layer? [1]

2. (a) A bottle in a laboratory has this symbol on the label.

   Why should you wear eye protection when using this substance? [1]

(b) Scientists use indicators to decide whether a substance is an acid or an alkali.

A student used the following colour chart to test different liquids.

Indicator	pH 1	pH 4	pH 7	pH 10	pH 13
Methyl orange	Red	Yellow	Yellow	Yellow	Yellow
Phenolphthalein	Colourless	Colourless	Colourless	Pink	Pink

Copy and complete the table, to show the missing indicator colours. [4]

Substance	Colour in indicator	
	Methyl orange	**Phenolphthalein**
Lemon juice	Yellow	Colourless
Vinegar	Red	
Bleach		Pink
Soap solution		Pink

(c) Which substance in the table is the strongest acid? [1]

(d) What further test would you need to do to identify which substance is the strongest alkali? [1]

3. Methane can exist in all three states of matter.

Draw a diagram to show the arrangement of particles in methane when it is a gas. [2]

4. Usman buys a new saucepan.

It is made of aluminium with a plastic handle.

(a) Suggest and explain *two* reasons why aluminium was chosen for the pan. [4]

(b) Suggest *one* reason why plastic was chosen for the handle. Explain the reason. [2]

Aluminium   Plastic

[total 23 marks]

Physics

Chapter 8

Energy

What's it all about?

How do you make things happen every day?
What do you do when you wake up? How do
you travel to school? How is your food cooked?
How do you light your rooms to see and read?

You can do some things using your own
energy. Understanding the physics of energy
enables you to make use of many energy
sources.

You will learn about:

- Different energy types
- Energy transfers
- Energy being transferred between different types
  but never lost
- How to make the best use of energy resources

You will build your skills in:

- Making careful observations and presenting results
- Finding information
- Applying science to daily life
- Making predictions and evaluations

Using energy

Learning outcomes
- To identify the different types of energy
- To describe energy transfers
- To recognise the sources of energy involved in different activities, including food and fuels

Starting point

You should know that...	You should be able to...
Energy can make things happen	Make careful observations
When materials get warmer, the particles (atoms and molecules) in them move around more	Present results in the form of a table

Activity 8.1: What do you know about energy?

Work in groups to find out how much you already know about energy.

- What is energy?
- What does energy do?
- Where does energy come from?
- How many types of energy can you name?

Write your answers on a board or a large sheet of paper so that everyone can see them.

Activity 8.2: Things that use energy

Try out several devices that make something happen. For example, devices that make a noise, produce light, or move. How is energy involved in each? Record your observations in a table like this:

8.1 *This clockwork toy moves when the energy stored in its spring is transferred to its wheels.*

Device	What it does	How it gets energy	Your observations and comments
Printer	Prints documents	From electricity	It also makes a noise and vibrates
Hole punch	Punches holes in paper	From push from hand	It makes a noise as it punches the holes

Moving

When you walk, you use energy to make you move. You got that energy from eating food. Motor vehicles move by using energy from the fuel that is burned in the engine. That energy is transferred into the energy of movement, called **kinetic energy**. The amount of kinetic energy an object has depends on its speed.

8.2 *Skilful snooker players know how much energy needs to be transferred to knock a ball into a nearby pocket.*

These transfers of energy to make things move are called **work**. The engine of a bus does work, making the vehicle move. Your legs do work when you walk. A snooker ball does work when it strikes another ball and makes it move.

Lifting or stretching

You use energy when you lift a heavy bag up onto a table. This transfer of energy from you to the bag is work. When you place your bag on the table top, it is not moving. Where has the kinetic energy of the moving bag gone?

The bag stores the energy because of its new position. We call this type of energy **potential energy**. If the bag is lifted higher, it stores more potential energy. You can use this stored potential energy to move something. For example, a cyclist at the top of a hill has stored potential energy. This is transferred into kinetic energy of the moving bicycle and rider as they go down the hill.

You can also store potential energy by stretching or squashing a springy material. For example, a stretched elastic band has potential energy.

Activity 8.3: Investigating energy transfers

Plan an experiment to investigate the motion of a wheeled trolley (or a model car) when you release it from different positions on a sloping runway.

A1 What can you observe?

A2 What can you measure?

A3 Design a table to record your results.

Key terms

kinetic energy: energy stored by an object because it is moving.

work: transfer of energy that causes an object to move.

8.3 *You can use the potential energy in a stretched elastic band in a catapult. The potential energy can be transferred to the kinetic energy of a small moving object.*

Key term

potential energy: the amount of stored energy something has because of its position.

Hydroelectricity

On a large river, a dam can be built to hold back a large lake of water. This lake stores potential energy. The water can be used to turn an electrical generator. The potential energy of the water is transferred to kinetic energy. This kinetic energy is transferred to electrical energy. China uses dams like this to produce about 17% of the electrical energy it needs for homes and industry.

8.4 *The Three Gorges Dam goes across the Yangtze River in China.*

1 State the type of energy a moving hammer has.

2 Wind turbines generate electricity from the energy of wind. State the type of energy wind has.

3 State the type of energy the water, held back behind a dam, has.

4 Look at figure 8.5. Copy and complete the sentences. Choose the best words from the list.

> speed   length   height   diameter

Object

Moving object

8.5

To work out kinetic energy, we need to measure the object's _____.

To work out potential energy, we need to measure the object's _____ above the ground.

5 Mia holds a ball at head height. She then drops the ball and it falls to the ground.

   a) Name the type of energy stored before the ball is dropped.

   b) Name the type of energy stored at the instant before the moving ball hits the ground.

   c) Describe the energy transfer that takes place.

6 Explain the energy transfers that happen in a hydroelectric power station. What type of energy does the water start with? To what type of energy is it transferred?

Energy sources

Your body's energy comes from digesting the food you eat. Vehicle engines get their energy by burning fuel. Digestion and burning are both chemical processes. Energy is transferred during those chemical reactions. We call it **chemical energy**.

A battery also stores chemical energy. If we connect the battery to a lamp (bulb) using wires, the chemical energy of the battery is transferred to **electrical energy**. In the lamp, the electrical energy is transferred to **light energy**.

8.6 *An electric car having its batteries charged.*

7 Describe the energy transfers involved in an electric car. What type of energy is stored by the battery? What type of energy is used to get the car moving? What type of energy does the car have when it is moving?

Solar panels use light energy from the Sun to produce electrical energy.

Sound also carries energy. A microphone transfers **sound energy** to electrical energy so that we can send voice messages by telephone or radio.

All substances are made of tiny particles (atoms or molecules). These particles move around or vibrate. We can measure the temperature of a substance. This tells us how much the particles in the substance are moving around or vibrating. The energy they store is called **thermal energy**.

If we put a hot substance near to a cold substance, energy is transferred from the hot substance to the cold substance. The thermal energy that is transferred is called **heat**.

8 Copy and complete these sentences. Use words from the box to fill the gaps.

> work, heat, chemical, light, sound, thermal

a) The heat from a flame comes from burning gas, which transfers _____ energy.

b) In an electric kettle, electrical energy is transferred to _____ energy to raise the water temperature.

Key terms

chemical energy: energy that can be transferred in a chemical reaction, e.g. burning fuel.

electrical energy: energy that can be transferred from a battery or power supply.

heat: thermal energy that is transferred from a hot object to a colder object.

light energy: energy that is transferred by a light source.

sound energy: energy that is transferred by a vibrating object making a noise, e.g. a musical instrument.

thermal energy: energy stored in an object due to its temperature.

c) _____ energy that travels through the air to our ears enables us to hear one another's voices.

d) Digesting food provides the _____ energy to enable us to do _____ like lifting or running; and it also provides _____ energy to keep our bodies warm.

Check your skills progress:

I can use a table to record my observations.

Energy conservation

Learning outcomes

- To understand that energy is always conserved: it cannot be created or destroyed
- To solve problems using the principle of conservation of energy
- To make careful observations

Starting point

You should know that...	You should be able to...
There are different types of energy	Use a table to record observations
Energy can be transferred from one type to another	

Pendulum

Activity 8.4: Observing the motion of a pendulum

Observe a pendulum.

A1 When is it highest and when is it lowest?

A2 When does it move fastest? Does it stop moving? If so, where?

A3 What kinds of energy does it have at each point in its motion?

Design a table to present your observations.

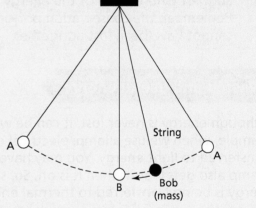

8.7 *A pendulum demonstrates the principle of energy conservation.*

A pendulum will swing backwards and forwards for a long time. Its motion demonstrates an important principle. Energy is not destroyed or lost. This is the principle of **conservation of energy**.

The pendulum's total energy stays the same all through the swing.

At point A, the mass is at its highest point. The mass has potential energy. At that instant, it has stopped moving upwards and is about to start moving downwards.

Key term

conservation of energy: energy cannot be created or destroyed. The total amount of energy is constant.

The mass swings back down and its speed increases.

At point B, the mass is at its lowest point. Its speed is at its largest value.

Potential energy has been transferred into kinetic energy.

Next, the mass begins to go higher again, but on the other side of the swing. Kinetic energy is being used to lift the mass. So, the kinetic energy transfers back into potential energy.

No energy is lost. The height at the end of each swing should be the same.

1 Look at the diagram shown in figure 8.7.

a) At a point somewhere between point A and point B, state the *two* types of energy that the mass has.

b) Describe what the principle of conservation of energy tells you about the total energy.

2 If a pendulum swings in a vacuum it might swing for ever. In air, the swings become gradually shorter. Suggest where some of the energy may have gone. Remember, the conservation principle means that it cannot have been lost altogether.

Wasted and useful energy

Although energy is never lost, it can be wasted. For example, when we use a lamp, electrical energy is transferred to light energy. You may have noticed that a lamp also gets warm when it is on. So, some electrical energy is being transferred to thermal energy.

The light energy is useful energy (it is what we use the lamp for).

The thermal energy is wasted energy.

3 In the example of the lamp, what does the conservation of energy tell us about the total of useful energy and wasted energy?

Key fact:

✔ Energy is conserved: it is never created or destroyed but just transferred from one energy type to another.

Check your skills progress:

I can make careful observations.

Describing energy transfers

Learning outcomes

- To use Sankey diagrams to explain the different energy transfers that take place
- To apply scientific knowledge and recognise how this affects our daily life

Starting point

You should know that...	You should be able to...
Energy is always conserved	Make careful observations
	Recognise energy transfers

Activity 8.5: Energy usage: past, present and future

Working in groups, discuss what energy is used for in the modern world. For example, lighting, carrying goods...

A1 What alternatives for each use (if any) were available 100 years ago?

A2 What other energy alternatives might become available in 100 years' time?

A3 Create a poster display that shows the changes over time in equipment and energy sources used.

Key term

joule: the scientific unit for energy. Its abbreviation is J. 1000 J = 1 kilojoule (kJ).

8.8 To keep warm in cold weather, you need food as your source of energy and thick clothing that reduces the amount of thermal energy transferred from you to the cold surrounding air.

Energy measurement

Energy transfers are measured in **joules** (J). A joule is quite a small unit, so we often use kilojoules (kJ). 1 kJ = 1000 J. You will often see those units used on food packaging. Food is the energy source for our bodies.

Energy transfer (Sankey) diagrams

Remember that energy is always conserved. The total energy that is transferred in must equal the total energy transferred out. Another way of saying this is:

total energy input = total energy output

We can show this in a diagram called a Sankey diagram (figure 8.9).

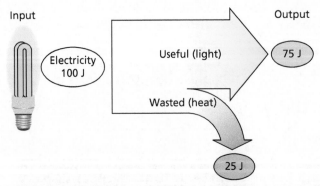

Input

Output

Electricity
100 J

Useful (light)

75 J

Wasted (heat)

25 J

8.9 *Energy transferred by a fluorescent lamp.*

In Sankey diagrams the width of the arrows shows the quantity of energy transferred. The energy input is shown on the left. Useful energy transfers are shown by horizontal arrows. Arrows pointing downwards represent wasted energy.

1 A car burns fuel with an energy value of 85 000 J. 68 000 J of this is output from the engine as wasted heat energy and sound energy. Calculate how much useful energy is left for moving the car.

2 Look at figure 8.9.

a) Write down the amount of energy that is useful.

b) Write down the total amount of energy that is input.

c) Explain what happens to the wasted energy.

3 For the fluorescent light bulb shown in figure 8.9, calculate the percentage of the input energy that is transferred usefully.

4 An old-fashioned filament lamp wastes 95% of the energy supplied to it. Draw an energy transfer (Sankey) diagram for it.

5 What percentage of the chemical fuel energy used by the car described in question 1 will *eventually* end up as thermal energy?

Activity 8.6: How energy makes things work

A1 Working in a small group choose four different devices that use energy. For example, you could choose a kettle, a clockwork toy, a television and a freezer.

A2 For each device, draw a Sankey diagram to show what type of energy goes in and what types of useful and wasted energy come out.

Key facts:

✔ The units of energy are joules and kilojoules.

✔ A Sankey energy transfer diagram can describe energy transfers.

✔ The amount of energy that has been wasted can be calculated from the difference between the total energy input and the useful outputs of energy.

Check your skills progress:

I can communicate my ideas and support them with evidence.

Finding enough energy

Learning outcomes

- To compare different energy resources
- To describe the differences between renewable and non-renewable resources
- To understand the importance of questions, evidence and explanations

Starting point

You should know that...	You should be able to...
The units of energy are joules and kilojoules	Communicate ideas supported by evidence
A Sankey energy transfer diagram can explain energy transfers	
The amount of energy that has been wasted can be calculated from the difference between the total energy input and the useful outputs of energy	
Plants require light and water for life and growth	

Energy resources

As time goes by, the number of people living on Earth increases. As we invent new devices, we also need more energy to power them. This means that we need more and more energy from the natural resources around us.

Today the large majority of our energy comes from coal, oil and natural gas. These are called **fossil fuels** because they come from the fossilised remains of organisms. These are a chemical store of energy that originally came from the sunlight which made the plants grow. Because it takes hundreds of millions of years for coal, oil and gas to be produced, they are **non-renewable energy resources**. If we keep burning them the supplies will run out.

Another non-renewable resource is nuclear fuel. You will learn more about nuclear energy in Stage 9.

8.10 *Fossilised tree fern leaf in a piece of coal: age about 300 million years.*

Key terms

fossil fuels: fuel, such as coal, oil and natural gas, made in the ground over millions of years from dead organisms.

non-renewable energy resources: resources that cannot be replaced quickly.

Renewable energy sources

Energy resources that can be replaced quickly are called **renewable energy resources**. Sunlight provides the thermal energy that causes all our weather. Sunlight also provides the light energy that plants use to grow. Many of the renewable resources we use are renewed by energy from the Sun.

But some renewable energy resources gain energy from other processes. The pull of the Moon's gravity creates tidal flows in the sea. Liquid rock in the Earth's mantle is kept hot mostly by the decay of the radioactive elements it contains.

Renewable energy	Energy source:	Renewed by:
Hydroelectricity	Stored potential energy of water in rivers and lakes	Rainfall over the year (Sun's energy)
Wind turbines	Movement of air – kinetic energy	Windy weather (Sun's energy)
Wave power	Up and down movement of water – kinetic energy	Windy weather (Sun's energy)
Solar panels	Sunlight – light energy	Sunny weather
Biomass and biofuels	Wood and other plant materials – chemical energy	Plant growth (Sun's energy)
Tidal power	Horizontal movement of sea water – kinetic energy	Tides due to the Moon's gravity
Geothermal energy	Hot rocks underground – thermal energy	Radioactive decay in underground rock

Table 8.1 *Examples of renewable energy*

Some of these renewable energy sources depend on the weather. The amount of energy they produce changes over time. So, we either use the energy they produce at the time they produce it, or we store the energy using large batteries.

 Explain how you could use electricity from solar panels, together with other renewable energy sources, to get the electrical energy you need at any time.

Energy from plants

Sunlight makes plants grow, so wood from trees can be a renewable fuel. The trees that are cut down must be replaced with new trees.

Plants other than trees can also be grown and then burned.

8.11 *Sugar cane is grown together with other fast-growing plants and used for fuel in India.*

The power of the sea

Island nations like the Philippines sometimes suffer from violent weather or from tidal waves. But we can also use the energy from the movement of the sea to produce electricity.

Energy for electricity generation

Electricity is a very useful form of energy. However, some methods of making it can cause a lot of pollution. For example, fossil fuels need to be collected from under the ground. Burning fossil fuels produces carbon dioxide and other waste gases.

Most large power stations use a source of thermal energy to boil water and produce steam. The heat can be transferred from the chemical energy of a fossil fuel or biomass. Heat can also be transferred from the nuclear energy of a nuclear fuel or from geothermal energy.

8.12 *China aims to reduce coal-burning to improve the air quality in its cities.*

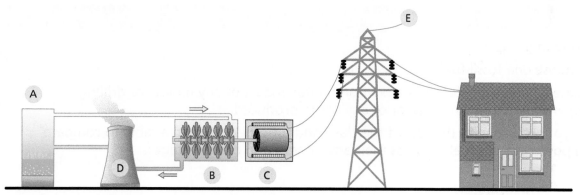

8.13 *The components of a power station: A – boiler, B – turbine, C – generator, D – cooling tower, E – wires to transfer electricity.*

2 Name *four* sources of chemical energy that can be used in power stations.

3 Copy and complete each of the following sentences with the correct words for energy type or energy transfer:

> *electrical energy   heat   kinetic energy   potential energy*

a) In the boiler, _____ is transferred to boil water and make fast-moving steam.

b) In the turbine, the steam transfers energy to turn the blades. The turning blades have _____ _____.

c) In the generator, the _____ _____ from the turbine is transferred to _____ _____.

d) In the cooling tower, the steam is cooled to form water that can be re-used. Thermal energy from the steam is transferred as _____ into the surroundings.

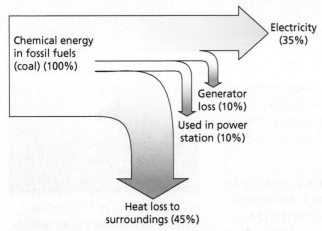

8.14 *A Sankey energy transfer diagram for a typical coal-fired power station.*

Activity 8.7: Fuels for power stations

Working in groups:

A1 Choose *one* fossil fuel.

A2 Research on the internet the environmental impact of any mining or drilling operations to get the fuel, or of any waste produced.

Share your results with the rest of the class, and together build up a table to compare these power station fuels. Discuss their advantages and disadvantages.

4 Which methods of generating electricity are renewable? List them, and explain why.

5 What problems are caused by burning large quantities of fossil fuels?

6 Hydroelectricity is clean to produce and relatively cheap. Suggest why it is not the main method of producing electrical energy in most countries.

Key facts:

✔ Fossil fuels took hundreds of millions of years to form underground, so are non-renewable.

✔ Burning fossil fuels produces waste gases that damage the environment.

✔ Renewable energy sources can be replaced quickly by natural processes.

Check your skills progress:

I can ask appropriate questions when choosing energy resources for generating electricity.

End of chapter review

Quick questions

1. What types of energy do the following objects have:

   (a) an arrow in flight? [1]

   (b) water at the very top of a waterfall? [1]

   (c) a pendulum mass at the centre of its swing? [1]

   (d) a stretched spring? [1]

   (e) a hammer lifted high? [1]

   (f) the same hammer just before it hits a nail? [1]

2. For each of the following, give the energy type involved:

   (a) fatty or sugary foods [1]

   (b) matches [1]

   (c) a battery [1]

   (d) sunshine [1]

   (e) a whistling noise [1]

   (f) hot steam. [1]

3. What kind of energy transfer happens during the following actions:

   (a) firing a catapult? [1]

   (b) a car or bicycle slowing down when its brakes are applied? [1]

   (c) lifting books to put them on a shelf? [1]

   (d) melting ice? [1]

4. How many joules of energy are there in a kilojoule? [1]

Connect your understanding:

5. Explain, using energy transfers, how a rubber ball bounces. [4]

6. Explain why heat or thermal energy cannot be seen. [2]

7. Heat is transferred in each of the following processes. State in which of them the heat is a useful energy output, and in which it is wasted:

   (a) cooking food in a microwave oven [1]

   (b) using an electric light. [1]

8. For the example in question 7 where the heat was wasted, explain what useful energy transfer also occurred. [1]

9. An electric motor in a vacuum cleaner uses 300 J of electrical energy every second. The energy wasted in the motor as heat is 60 J every second.

   (a) How much useful energy is transferred every second? [2]

   (b) Draw and label a Sankey diagram for the vacuum cleaner. The labels should state the amounts and types of energy being inputted and outputted. [5]

10. What factors should a government consider when planning a new electricity generating plant? [4]

11. A pole-vaulter's muscles are not very efficient at transferring energy into useful work.

    (a) What *two* kinds of useful energy does a pole-vaulter need when making a jump? [2]

    (b) What type of energy is stored in the food an athlete eats? [1]

    (c) What form does the wasted energy take, and how can you observe it? [2]

    (d) Draw and label a Sankey energy transfer diagram to represent the energy transfers involved in a pole-vault. [5]

Challenge question

12. When the vacuum cleaner from question 9 is used inside a closed room, the air in the room gradually gets warmer.

    (a) Describe the energy transfers that take place to cause this warming. [4]

    (b) Explain why all the electrical energy input will eventually end up as increased thermal energy in the air. [2]

Chapter 9

Beyond the Earth

What's it all about?

Looking into the sky, you can see objects beyond the Earth – the Sun, the Moon, stars and planets. Find out how astronomers like Copernicus and Galileo explored beyond the Earth without travelling into space. What have we discovered out there since, and where might we travel in the future?

You will learn about:
- Why we have day and night, seasons and years
- The Moon, and how its movement explains eclipses
- The other planets in the Solar System, and how they compare with the Earth
- Our Sun and other stars: how we see them and how they help us to see other things in space
- How our understanding of the Solar System has developed, and that it is still developing

You will build your skills in:
- Finding and using information
- Recognising patterns in observations
- Talking about questions, evidence and explanations

Day and night

Learning outcomes
- To describe how the movement of the Earth explains the apparent movement of the Sun and the stars
- To define a day, a month and a year
- To explain why we experience seasons on Earth

Starting point

You should know that...	You should be able to...
We see the Sun rise and set each day and that the Moon, stars and planets can be seen at night	Use the points of the compass – north, east, south, west

Activity 9.1: What is moving?

A In groups, find a safe way to make one member of the group spin, for example by standing on tiptoe, or by sitting on a seat that revolves. Make sure that the others in your group stand clear, but that you are all watching and ready to catch or support the spinning person. Only do it for a short time, and take turns to spin.

B Discuss the differences between your observations when spinning and those when you were watching. How do the room and the people in it appear?

A1 Look at the three pictures in figures 9.1 to 9.3. In each, what is moving? How did you decide?

9.1 *A bus moving.*

9.2 *Running animals.*

9.3 *Stars in the evening sky.*

The spinning motion of the Earth

When you look up at **stars** in the night sky, they seem fixed in their positions. But, if you look again an hour later, they will have moved. In the **northern hemisphere**, there is just one star that seems to stay fixed. It is the star that seems to be above the Earth's north **pole** – it is called the North Star (or sometimes the Pole Star or Polaris).

Key terms

northern hemisphere: the half of the Earth's spherical surface that lies north of the equator.

star: huge mass of gas that is undergoing nuclear reactions. Stars are so hot that they give out light all the time.

Look at the photograph of the North Star in Figure 9.4. It looks as if all the other stars are spinning round it. But that is an illusion. It is the Earth that is spinning.

The Earth spins on its **axis** once every 24 hours. This explains why the stars seem to rotate at night. It also explains why the Sun seems to move across the sky each day, rising in the east and setting in the west.

9.4 *A photograph of the North Star and neighbouring stars, taken over a period of 45 minutes.*

9.5 *The Earth is a sphere, spinning on a tilted axis.*

When you stand on Earth, you feel as if you are still. But the nearer you are to the **equator**, the faster you are moving. At the equator, you are spinning at over 1500 km/h. Standing at one of the poles, you turn slowly around on the spot.

1 How long does it take you to rotate 360°: **a)** if you stand at a pole? **b)** if you are on the equator?

2 Explain why the North Star stands still, while all the other stars seem to move.

Key terms

axis: imaginary straight line, running through the centre of the Earth between the two poles.

equator: imaginary line running around the circumference of the Earth halfway between the north pole and the south pole.

pole: the north and south poles mark the ends of the axis about which the Earth spins.

Earth's orbit around the Sun

As well as spinning on its axis, the Earth also makes an almost circular journey – an **orbit** – round the Sun every year. Our Sun is a star 150 million kilometres away from us, and so the Earth is travelling very quickly – over 100 000 km/h. Gravity keeps the Earth in this orbit. You will learn more about this in Chapter 10.

The Earth spinning on its axis causes the positions of the stars to appear to rotate during the night (see figure 9.4). The Earth orbiting the Sun causes the position of stars to appear to move over a year.

Key term

orbit: circular or elliptical path in which one object travels round another. Gravity holds the objects together.

Daylight hours and the seasons

The Earth's axis tilts compared to the way it orbits the Sun. The spin axis has a **tilt** of 23°. This is why the length of the day changes with the **seasons** of the year.

This side of the Earth is facing towards the Sun: it is day here
This side of the Earth is facing away from the Sun: it is night here

23°
Summer in northern hemisphere
Winter in the southern hemisphere

Winter in northern hemisphere
Summer in the southern hemisphere

Earth rotates on its axis once every 24 hours

Earth orbits the Sun once a year

9.6 *The Earth's tilt means that the length of day and night are not always the same.*

In figure 9.6 (right side), the north pole is tilted away from the Sun and the south pole towards the Sun. So, places north of the equator get a longer time in darkness (night) and a shorter day. They are experiencing the season of winter. Places in the southern hemisphere get longer days and shorter nights, so there it is summer.

Activity 9.2: Modelling Earth's orbit and changing day and night

Work in groups in a partly darkened room. Use a torch and a foam ball with a thin rod stuck through it to model the Earth's spin and its orbit around the Sun. Keep the torch pointing towards the ball all the time. Use your model to answer the following questions:

A1 Where in the sky is the Sun at midday:

    a) if you live north of the equator?

    b) if you live in the southern hemisphere?

A2 How many times does the Earth spin round during one orbit of the Sun?

3 Explain why the stars that you see from a place in the southern hemisphere are different from those seen from the northern hemisphere. (Imagine stars on the ceiling and on the floor of the room.)

Key terms

seasons: divide the year into four periods of three months each: winter, spring, summer and autumn (or fall). Winter has shortened daylight hours and summer has the longest. On the equator, day and night are always 12 hours each, so there are no seasons there.

tilt: the angle between the Earth's spin axis and the axis of its orbit around the Sun.

The Moon's orbit

The largest and brightest object in the night sky is our Moon. Like the Earth, it is a sphere and is lit by the Sun, so it is bright on one side and dark on the other side. It orbits the Earth once every 27.3 days.

How much we can see depends on the positions of the Moon and Earth compared to the Sun. This is why the appearance of the Moon changes (see figure 9.7).

9.7 *From one New Moon (completely dark) to the next takes 29.5 days – one lunar month.*

4 Copy and complete the following sentences. Use words from the box to fill the gaps:

> day, Earth, month, night, orbit, Sun

a) The _____ and the Moon are both lit by the _____ .

b) The length of _____ and _____ on Earth always totals 24 hours.

c) A year is the time it takes the _____ to orbit the _____ .

d) The length of one lunar 'day' and 'night' on the Moon add up to one lunar _____ .

Key facts:

✔ The Earth spins round once on its axis every day.

✔ The Earth orbits the Sun once each year.

✔ The Earth's spin axis is tilted.

Check your skills progress:

I can link the seasons to the changing length of daylight time.

I can explain the apparent movement of the stars.

I can use a model to explain day and night, and the orbit of the Earth around the Sun.

Planets and the Solar System

Learning outcomes

- To name the planets in our Solar System
- To describe how the Sun and other stars are sources of light
- To describe what happens in solar and lunar eclipses
- To describe how planets and other bodies are seen by reflected light
- To describe how the planets move around the Sun
- To explain why a day and a year on another planet are different from those on Earth

Starting point

You should know that...	You should be able to...
We can see things if they give out light or if they reflect light	Find information from books or using the internet

Planets and their moons

After our Moon, the largest and brightest objects in the night sky are **planets**. They do not always appear next to the same stars. Through our year, they seem to wander across the sky. Their paths appear to us as complicated looped shapes.

Like the Earth, each planet is in an orbit around the Sun. Including the Earth, the Sun has eight planets orbiting it, each at a different distance (see figure 9.8). Each planet has its own 'year' (the time for one orbit round the Sun) and 'day' (the time it takes to spin once on its axis).

If you observe Jupiter using a telescope, you can see it has **moons** circling it. In fact, many of the planets have moons of their own. If you observe Saturn you can see it also has rings (made of thousands of small icy objects).

There are other objects orbiting the Sun that are not planets. Pluto is a dwarf planet that spends most of its orbit outside the orbit of Neptune. Beyond Pluto are icy objects called comets. Between Mars and Jupiter are four more dwarf planets and smaller rocky objects called asteroids.

This collection of objects moving around the Sun is called the **Solar System**.

Key terms

moon: natural object made of rock or frozen liquid, which orbits a planet.

planet: object which orbits around the Sun and is large and heavy enough to have become approximately spherical and to have cleared all other smaller bodies out of its orbit.

Solar System: the Sun and all the other objects that move around it under the control of its gravity.

9.8 *The eight planets of the Solar System (distances from the Sun and the sizes of the planets are not drawn to scale). You should learn and remember their names and positions.*

	Mercury	Venus	Earth	Mars	Jupiter	Saturn	Uranus	Neptune
Mass (10^{24} kg)	0.330	4.87	5.97	0.642	1898	568	86.8	102
Rotation period (hours)	1408	5833	23.9	24.6	9.9	10.7	17.2	16.1
Length of day (hours)	4223	2803	24.0	24.7	9.9	10.7	17.2	16.1
Distance from Sun (millions of km)	58	108	150	228	779	1434	2873	4495
Orbit period (Earth days)	88.0	224.7	365.2	687	4331	10 747	30 589	59 800
Average surface temperature (°C)	167	464	15	−65	−110	−140	−195	−200
Atmospheric pressure at surface (kPa)	0	9200	100	1	unknown	unknown	unknown	unknown
Number of moons	0	0	1	2	67	62	27	14

Table 9.1 *Planet data*

1. Table 9.1 includes information about the average surface temperature of the planets. Explain which other line of data best explains the pattern of average temperatures at each planet's surface.

2. Again, referring to data in Table 9.1, notice that for some planets their rotation period is very different from their length of day. Explain for which planets this happens and draw a diagram to show why.

3 Which planets are closest to the Earth's orbit?

4 Which planet has the longest day? Explain how that fits a general pattern. Which planet is an exception to this general pattern?

5 Suggest why Pluto does not fully fit the modern definition of a planet.

How you see things

You can see a source of light, for example an electric lamp. You can also see things that reflect light, for example a page from this book. You cannot read a book without a light source shining on it.

Stars are the only light sources in space. Our Sun is a star – the closest star to Earth. The next nearest stars are hundreds of millions of kilometres away, so they look small and their light is faint. You can only see planets, moons or other objects such as asteroids if they reflect light from the Sun.

Some astronomers think there may be a ninth planet that has not been discovered because it is so far from the Sun that it does not get enough light to show up in telescopes.

Activity 9.3: What you can see – and what you cannot

In groups:

A1 Discuss why you only see stars other than the Sun at night.

A2 Discuss why it is easier to see the bigger planets that are close to the Earth, but harder to see the smaller objects that are further away.

A3 Draw a labelled diagram to explain how you can sometimes see the Moon during the day.

Eclipses – darkening of the Sun or Moon

The Moon's orbit can sometimes cause an **eclipse**. This happens when the Sun, the Earth and the Moon all fall into a straight line.

Warning: Never look directly at the Sun. It can seriously damage your eyes even during an eclipse.

Key term

eclipse: when one object interferes with our view of another object.

Solar eclipse	Lunar eclipse
During the day, the Moon gets right between Earth and the Sun, so our view of the Sun can be blocked out.	At night, the Earth gets between the Sun and the Moon; you can see the Earth's shadow pass slowly across the Moon, darkening it.
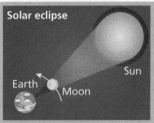 9.9 *Diagram to show a solar eclipse.*	 9.10 *Diagram to show a lunar eclipse.*
9.11 *In a solar eclipse the Moon blocks our view of the Sun.*	9.12 *In a lunar eclipse the Earth casts a shadow over the Moon.*

Table 9.2 *Comparing solar and lunar eclipses.*

Activity 9.4: Solar and lunar eclipses

In groups, demonstrate eclipses. Use a tennis ball to represent the Moon, a football to represent the Earth, and a torch light to represent the Sun.

A1 Look towards the torch and move the tennis ball to see 'the Moon' eclipse 'the Sun'.

A2 In a darkened room, place a football (the Earth) between the torch (the Sun) and a tennis ball (the Moon) to make a lunar eclipse.

A3 Use notes and drawings to describe the different appearance of solar and lunar eclipses.

Key facts:

✔ Eight planets, including the Earth, orbit our Sun.

✔ Each planet has its own orbit and spin, giving it a different 'year' and 'day' from the Earth's.

✔ We can see stars (including the Sun) because they produce their own light.

✔ We can see planets and moons because they reflect the light of the Sun.

✔ When the Moon passes in front of the Sun, we see a solar eclipse.

✔ When the Earth passes between the Sun and the Moon, we see a lunar eclipse.

Check your skills progress:

I can name the eight planets, and place their orbits in order of distance from the Sun.

I can explain what defines a planet.

I can find and interpret information about the Solar System from secondary sources.

I can recognise patterns in information and identify gaps or unexpected values.

I can explain how solar and lunar eclipses occur.

Changing ideas about the Solar System

Learning outcomes

- To describe how the planets were discovered
- To discuss how we have changed our ideas about the Solar System because of scientific evidence
- To investigate the ideas of Copernicus and Galileo
- To be able to talk about the importance of questions, evidence and explanations

Ancient ideas challenged

For thousands of years, people have been looking up into the sky and wondering about all the lights they see. Many ancient cultures described the Earth as a flat disc with a domed sky above it.

By 330 BC the ancient Greek philosopher and scientist, Aristotle, had stated that Earth is a sphere. His evidence included sea travellers' observations that ships disappear over the horizon and that the North Star rises higher in the sky as you travel northwards. Observation of lunar eclipses showed that the Earth's shadow is always circular.

However, he thought that the stars, planets, Moon and Sun must all revolve around the Earth on spheres of different diameters.

9.13 *The Earth imagined as a flat disc floating on a sea with the stars on a spherical dome.*

Careful observations, new techniques, new evidence, new thinking

Astronomy continued to develop in other parts of the world, especially in India. In the sixth century, the Indian astronomer Arabyhata published an important book in which he said that the Earth spins and so causes the apparent motion of the stars at night.

Next, Islamic astronomers in Persia and Arabia built on the Greek and Indian writings. They developed new mathematics and better methods of observation. Between the tenth and the fifteenth centuries their observations included the first sighting of another **galaxy**, Andromeda, and recording the brightest supernova (exploding star) in history. They also invented algebra and a new physics of astronomy. In the sixteenth century, people reading about this Islamic work sparked a new interest in astronomy among Europeans.

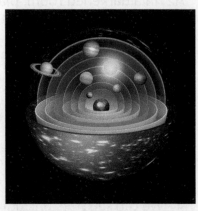

9.14 *Aristotle's Earth – a sphere at the centre of other revolving spheres.*

New theories proposed and tested

In 1543, a Polish priest called Nicolaus Copernicus published a new **theory**. He said 'We revolve around the Sun like any other planet'. His ideas were attacked, and some people were even killed for teaching them. After the invention of telescopes, Galileo Galilei used one in 1610 to look at Jupiter. He observed moons circling round the planet. His observations also proved that Venus orbits the Sun.

The mathematicians Johannes Kepler and Isaac Newton used the new astronomical observations to develop new **scientific laws** of motion and gravity. Kepler showed that the orbits of planets are not perfect circles but ellipses – their distance from the Sun has a maximum and a minimum point. Newton explained this, using his new laws of motion and of gravity, which were first published in 1687. Newton's laws changed the whole basis of physics.

9.15 *An Arabic astrolabe – an instrument first invented by Greeks that measures the angular position of an object in the sky.*

More planets predicted and found

Using telescopes to make accurate measurements of the planets' movement, scientists applied and tested Newton's laws of motion and gravity. The motions of the outer planets were not quite as predicted. They calculated that there must be other objects exerting gravitational forces. They searched for these using bigger telescopes and found two more planets – Uranus and Neptune. Finally, in 1930, the dwarf planet Pluto was first seen.

Exploring the Sun and the stars

At the beginning of the nineteenth century, the German scientist, Joseph von Fraunhofer, discovered dark lines in the spectrum of light coming from the Sun. These spectral lines have enabled us to discover what chemical elements are in the Sun and in other stars.

Early in the twentieth century, the American scientist, Annie Jump Cannon, used spectral data to classify more than 400 000 stars. This was more than any other scientist. Another member of her team, Henrietta Swan Levitt, discovered a type of 'variable' star that made it possible for us to reliably measure the huge distances in space between stars and ourselves.

Later, Russian-born American, George Gamow, was the first to put forward the idea that the energy of our Sun and of the other stars comes from nuclear fusion reactions.

Key terms

galaxy: collection of billions of stars held together by gravity. Our Sun with its Solar System is part of the Milky Way galaxy.

scientific law: a theory that has been tested by experiment and shown to be useful and reliable.

theory: a set of ideas that describes how things work.

Find out more about the scientists after Copernicus and Galileo who changed our ideas about the Earth and its place in the Solar System. Include those mentioned on page 185 and others you think are important.

A1 Draw a timeline showing each important new idea and the person who published it.

A2 What kinds of evidence did each scientist use?

International collaboration

The huge cost of space research means that much more can be achieved when nations work together; an example of this is the International Space Station (shared by the USA, Russia, Europe, Japan and Canada) and the Hubble Space Telescope (an ESA and NASA project). Photographs taken from these have changed the way humans think about their place in the universe. We are not at the centre of things, but on a tiny planet which orbits one star out of billions in our galaxy. Our galaxy is just one of an uncountable number that stretch as far as we can see.

9.16 *Earth viewed from the International Space Station.*

9.17 *The Hubble telescope ultra-deep field view shows that the 'black' areas in our night sky are actually full of hugely distant and very faint galaxies.*

1 What did Copernicus say about the movement of the Earth and Sun that disagreed with Aristotle's theory?

2 Describe the new instrument that Galileo used.

3 Kepler used many thousands of observations to develop his equations. Suggest why this use of evidence was so important.

Activity 9.6: Challenging ideas

In small groups, discuss what scientists need to do for a new theory to be accepted.

A1 Why are new theories needed, and how do they arise?

A2 If two theories could explain the same observations, how would you decide which one to accept?

Key facts:

✔ Aristotle showed from observations that the Earth is spherical.

✔ Copernicus proposed that the Earth is a planet orbiting the Sun.

✔ Galileo used a telescope to make more accurate observations of the planets and their motion.

✔ Kepler developed equations that described the orbits.

✔ Newton explained the orbits with simple mathematical laws of motion and gravity.

Check your skills progress:

I can explain how scientists used evidence to support new ideas about the Solar System.

I can give examples of how new theories are tested before they become accepted.

End of chapter review

Quick questions

1. How many days are in a lunar month? [1]

2. How many planets orbit our Sun? [1]

3. How far is the Earth from the Sun? [1]

4. Which planet orbits closest to the Sun? [1]

5. Which planet is furthest from the Sun? [1]

6. Which planet is largest? [1]

7. Which planet has the most moons? [1]

8. Look at the data in the table. Draw a diagram to explain why the length of the longest day in summer and the shortest day in winter is different in these three capital cities.

City	Rekyavik, Iceland	Tokyo, Japan	Nairobi, Kenya
Longest day	21 h 45 min	14 h 35 min	12 h 12min
Shortest day	4 h 7 min	9 h 44 min	12 h 2min
Longitude	21.8° W	139.7° E	36.8° E
Latitude	64.1° N	35.7° N	1.3° S
Distance from the equator	7115 km	3963 km	144 km

Connect your understanding

9. Which *two* planets were not known at the time of Copernicus and Galileo? Explain why. [3]

10. Give *three* connected reasons why Pluto is no longer counted as a planet. [3]

11. Explain how the Moon is lit and why it appears to change shape. [4]

12. Describe and explain the differences between a solar eclipse and a lunar eclipse. [7]

Challenge question

13. Explain why we only ever see one side of our Moon. [2]

10

Chapter 10
Forces and their effects

What's it all about?

We use the idea of energy (see Chapter 8) to describe what makes things happen. But although we know a golfer hitting a ball is transferring energy, that does not tell us the direction the ball moves in or how fast it moves. To understand these things, we need to talk about forces. The golfer uses a force with a size that explains how fast the ball travels. The direction of the force tells us the direction the ball moves in.

You will learn about:
- Changing an object's speed or its direction of travel
- Balancing forces, like friction or air resistance
- Gravity, free fall and orbital motion

You will build your skills in:
- Making and recording measurements
- Planning and carrying out investigations

Forces change motion

Starting point

You should know that...	You should be able to...
Forces are pushes or pulls and are measured in newtons (N)	Measure forces using a newton meter (force meter)
Motion has a speed and a direction	Present results in the form of a table

If something is not moving, but you want it to start moving, you need to give it a push or a pull. Imagine a shopping trolley on wheels. When you visit a supermarket, you go to where the trolleys are kept and push one trolley to make it move. In physics, we call the push a force.

As you push the trolley round the store, you speed up or slow down. You change the size of the force. You also change the direction of the trolley with pushes or pulls. You change the direction of the force to make this happen.

In 1687, Isaac Newton published this idea:

'An object will carry on moving with a steady speed in the same direction unless a force acts on it.'

Activity 10.1: Slowing down

On Earth, moving objects always seem to slow down and eventually stop. In groups, plan an investigation into the movement of different objects on different surfaces to find out which will travel the furthest before stopping. Consider:

- What surfaces to use. For example, a smooth dry surface, a wet surface, a rough surface...
- What objects to test. For example, something light, something heavy, something with wheels, something round...
- How to make a fair test.

A1 Then do your investigation. Record your results in a table.

A2 Explain how the stopping forces affected the distance each object travelled.

Showing forces on diagrams

You can show forces as arrows on diagrams. This helps you understand what forces are acting on objects and what these forces will do.

The rules for drawing forces are:

- The direction of the arrow shows the direction of the force.
- The length of the arrow is to scale, to show the size of the force.

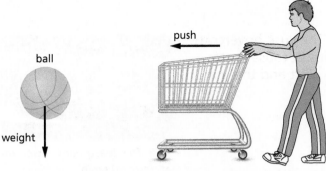

10.1 *Drawing diagrams helps us to understand forces. The weight of a ball causes it to speed up in a downwards direction. The push on a trolley causes it to speed up in a horizontal direction.*

Friction

When two solid surfaces rub together, **friction** between them acts to slow down and stop the movement. Friction between your shoe and the floor is very useful (see figure 10.3). It stops you slipping. But friction between the moving parts in an engine or a machine is a nuisance. We reduce it by using oil.

Air resistance

Gases and liquids provide some resistance to motion. If you ride a bicycle fast you will experience **air resistance**. It feels as if a wind is pushing you backwards, until you slow down and stop. This is a force that depends on the speed at which you travel.

The size of the force due to air resistance also depends on the size and shape of the moving object. Smaller, more smooth and rounded objects help to reduce the air resistance.

Activity 10.2: Investigating friction

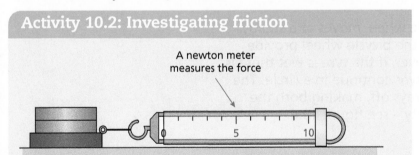

10.4 *Measuring the frictional force between a flat surface and a wooden block that has masses placed on it.*

Working in small groups, design, and then do, an investigation into the factors that affect the size of the frictional force on a wooden block.

A1 What is the effect on the friction force of increasing the weight that pushes the surfaces together?

A2 What is the effect on friction of different surfaces, or adding water or oil?

10.2 *There is friction between the tyres on a bicycle and the rough surface of the track. It means the tyres grip the track so the cyclist can pedal the bicycle forwards. The cyclist has to keep pedalling, or air resistance will slow the bicycle down.*

10.3 *A badminton player wears shoes with a good friction grip. Friction forces oppose the push of his feet and stop him slipping. After it leaves the racket, the shuttlecock slows down rapidly due to air resistance.*

Key terms

air resistance: the force that acts to slow down an object moving through air. It varies with the size and shape of the object.

friction: force between two surfaces that are pressed together. It acts to stop the surfaces sliding over one another.

Indian coastguard

India has a long coastline to protect. They use hovercraft, which can travel much faster than ships because an air cushion reduces friction between the craft and the sea or land below.

10.5 *This coastguard vessel travels at up to 90 km/h.*

Moving in circles

Activity 10.3: Turning corners

A Ensure you are wearing eye protection. Tie a cotton reel or a rubber bung to one end of a piece of string. Hold the other end and swing the object round in a circle. You can feel the pull in the string. That is the force needed to keep the object moving in a circular path.

B Now find a safe space (outside in an open area), where throwing the object cannot damage anything. Once again, swing the object round in a circle, but then let go of the string. In which direction does it fly off? Draw a diagram to show its flight.

If you want to change direction when you are already moving, you need to use a sideways force. For example, if you are turning left, you need a force pulling you to the left.

If a constant force acts pulling you to turn, and you stay at a constant speed, then you can turn in a circle. This force points towards the centre of the circle. It is called the centripetal force.

For example, the tyre of a bicycle wheel moves at a steady speed, in a circle. The spokes of the bicycle wheel provide the turning force needed. However, if the tyre is wet there is no force to make water on the tyre continue in a circle. The water carries straight on and sprays off, making both the rider and passers-by wet and dirty – see figure 10.6.

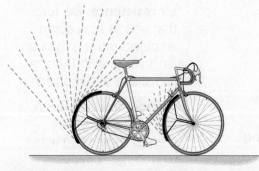

10.6 *Water on a spinning wheel tends to fly off in a straight line, so bicycles have mudguards (fenders).*

1 Describe what might happen if a car tries to turn a corner too quickly on a slippery road surface.

2 Draw and label diagrams to show the direction of the frictional force needed from the road acting on a bicycle's tyres:

   a) when the cyclist pedals hard to go faster

   b) when the cyclist turns a corner

   c) when the cyclist uses both brakes to stop quickly.

3 What would happen to the motion of objects moving in a straight line if there was no friction or air resistance?

Key facts:

✔ With no force, a stationary object stays still.

✔ With no force, a moving object continues at constant speed in a straight line.

✔ If a force is applied to an object it changes speed or direction.

✔ Turning a corner requires a sideways (centripetal) force.

✔ A force acting forwards speeds an object up; a backwards force slows an object down.

✔ Friction is a force between surfaces that slows things down.

Check your skills progress:

I can measure forces using a newton meter.

Gravity

Starting point

You should know that...	You should be able to...
Mass and weight are different and have different units	Measure forces using a newton meter
Forces can change the shape of an object	Record your results using a table

Free fall

Activity 10.4: Weight and free fall

A Hold different masses in your hand and feel the weight of each one.

B Weight is a force. Use a newton meter to measure the weight of each mass in newtons. Record your results in a table listing mass (kg) and weight (N).

A1 For each pair of readings in your table, calculate weight ÷ mass. What do you notice?

C Drop a heavy mass and a light mass into a cardboard box or tray on the floor at the same time and from the same height.

A2 What do you notice about the time it takes for the heavy mass and light mass to reach the ground?

Gravity is the pull force between two masses. The bigger the masses and the closer they are together, the bigger the attractive force between them. The Earth has a very large mass indeed, so all other objects near it experience a gravitational pull that we call **weight**.

1 Mars has a mass nine times smaller than the Earth's.

a) If you measured your mass on Mars, would it be less than, the same as, or more than your mass on Earth?

b) If you measured your weight on Mars, would it be less than, the same as, or more than your weight on Earth?

2 You are in an aeroplane flying high up in the sky and you measure your weight. What result do you expect? Explain why.

Key terms

gravity: the pull of a large mass like the Earth on other masses near it.

weight: the size of the pull of gravity for a given mass. You can feel something's weight when you try to stop it from falling.

The weight of an object is proportional to its mass.

Weight = $m \times g$

The 'g' is the strength of gravity. Approximately, for objects near the Earth's surface, $g = 10$ newtons per kilogram of mass.

This force makes objects fall towards the Earth. When objects fall in this way because of their weight, it is called **free fall**.

In the sixteenth century, Galileo did an experiment where he dropped objects of different mass from the Leaning Tower in Pisa, Italy. He found that they both landed at the same time.

Both objects gain speed at the same rate because the force needed to change an object's speed also depends on its mass.

3 Galileo's experiment only works with two dense objects. Explain why the result would be different if you dropped a hammer and a feather together.

Orbital motion

In Chapter 9 you found out about the Earth and other planets orbiting the Sun. The force that keeps the planets in their orbits is gravity – the pull of gravity from the Sun. The same is true for anything that orbits. For example, the Moon orbits the Earth due to the pull of gravity from the Earth.

Gravity is a force that acts at a distance. There is no string or spoke fixing these two masses together; there is just a force. So, the object's speed of travel around the circle must be carefully balanced against the pull of gravity (its weight). Things only stay in orbit if they are circling at the correct speed.

Pairs of forces

Through studying gravity, Newton realised that forces always seem to come in pairs that act in opposite directions. For example, the pull of Earth's gravity on the Moon keeps it in its orbit, while the pull of the Moon's gravity on the Earth makes the oceans move and so causes tides.

Another example is when we try to stretch or squash materials. You apply a force to an elastic band to make it stretch, and the elastic band pulls back on you. You can feel its force. Again, these two forces act in opposite directions.

10.7 *Galileo dropped balls of different sizes from the Leaning Tower.*

Key term

free fall: the speeding up downwards motion of a body due to its weight.

10.8 *Astronauts in a space station appear to be 'weightless'. But in fact their weight – the force of gravity from the Earth – is what is keeping them in the same circular orbit as the space station.*

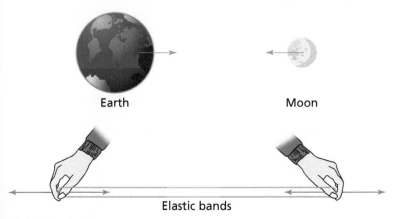
Earth   Moon

Elastic bands

10.9 *Pairs of forces.*

Activity 10.5: Changing shape

Newton meters work by stretching a spring. The force you apply makes the spring longer, and so the pointer moves along the scale.

A Make your own newton meter using an elastic band or a soft spring, with a ruler to measure its change in length. You could attach a needle to the moving end to act as a pointer.

B Decide what measurements and calculations you will need to make, and design a suitable table for recording your results. Remember, it is the *change* in length that you need to know.

C Design and carry out an investigation similar to the one in Activity 10.4, where you hang masses on your newton meter and measure the effect of the force.

   A1 How can you use your results from Activity 10.4 to convert the values of mass into weights measured in newtons?

D Investigate what happens to your elastic band or spring when you start to use heavier weights.

   A2 Is the change in length always proportional to the force applied? Explain why.

Key facts:

✔ The weight of an object on Earth is the pull force of the Earth's gravity on it.

✔ Weight is measured in newtons and is proportional to the mass of an object (in kilograms).

✔ Gravitational pull changes with distance from the Earth.

✔ Gravity is the force that explains the orbits of planets, moons and satellites.

Check your skills progress:

I can identify pairs of forces acting in opposing directions.

Balanced and unbalanced forces

Learning outcomes
- To describe the effects of forces on motion, including friction and air resistance
- To describe the effect of gravity on objects

Starting point

You should know that...	You should be able to...
Mass and weight are different and have different units	Identify pairs of forces acting in opposite directions
With no force, a stationary object stays still	
With no force, a moving object continues at constant speed in a straight line	
If a force is applied to an object it changes speed or direction	

Adding forces

Sometimes more than one force acts on an object. You can add these forces together to make a single force, called the **resultant force**. This resultant force shows you the combined effect of all the different forces on an object.

You have seen how to draw forces on diagrams. Drawing a diagram helps show how forces add up.

Key terms

balanced forces: when the resultant force is zero.

resultant force: shows the single total force acting on an object when all the forces acting on it are added up.

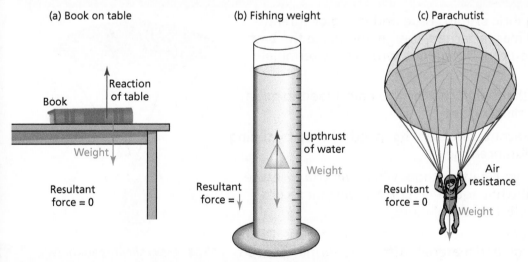

(a) Book on table

Reaction of table
Book
Weight
Resultant force = 0

(b) Fishing weight

Upthrust of water
Weight
Resultant force = ↓

(c) Parachutist

Air resistance
Resultant force = 0
Weight

10.10 *Adding forces and finding the resultant force.*

Sometimes the resultant force is zero. When this happens, we say that all the forces on an object are **balanced forces**.

So, that object either stays at rest, or if it is moving already it keeps a constant speed and direction. To balance, the two forces need to be equal in size but opposite in direction.

10.11 *These sky-divers will need to use a parachute to slow down.*

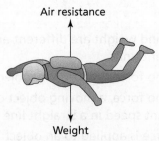

Air resistance

Weight

10.12 *Balanced forces result in a constant speed of fall.*

When the sky-divers in figure 10.11 jump out of a plane, they fall faster and faster. The force due to gravity makes them speed up. The faster they fall, the larger the force due to air resistance. After a time, their speed stops increasing and stays constant. This speed is called the **terminal velocity** (or terminal speed). It happens when the force due to air resistance slowing them down balances their weight (see figure 10.12).

This terminal velocity is much too fast for a safe landing, so they use a parachute to increase the air resistance force. This slows the sky-diver to a new, slower terminal velocity.

Key term

terminal velocity: the steady speed of an object with balanced pull and air resistance forces acting on it.

Activity 10.6: Make a parachute

A Using light fabric or thin tissue and string or thread with a small mass attached, design and make the best parachute you can. Design one with a slow terminal velocity.

  A1 Explain the design choices you have made to get a slow terminal velocity.

B Test your parachute. Measure its speed of drop by timing it. Speed = distance/time.

  A2 Describe your test method. What will you measure and with what equipment? How will you record your results and calculations?

C Compare your results with those of other groups. Discuss how the design and materials affect the terminal speed.

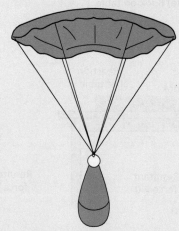

10.13 *A very simple parachute.*

Floating

When an object floats in water, the water pushes upwards and balances the weight of the object. The upwards force from the water is called **upthrust**.

Activity 10.7: Measuring upthrust

A Hang a wooden block from a newton meter and measure its weight.

B Slowly lower the block into a beaker of water. What do you notice about the reading?

  A1 Describe and explain how the upthrust on the wooden block changes as the block is lowered into the water.

C Repeat the investigation using a metal object – something that will not float – instead of the wooden block. Do you observe an upthrust on the metal object? If so, how large is it?

  A2 Explain, using the idea of balanced or unbalanced forces, why the metal object cannot float.

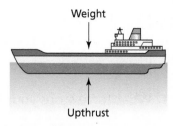

Weight

Upthrust

10.14 *The forces that allow a ship to float.*

Key term

upthrust: the upwards force from a liquid on a wholly or partly submerged object.

---

1 A bus with a weight of 400 000 N crosses a bridge. What upwards force must the bridge structure be able to provide?

2 Because of their shape, an aeroplane's wings cause an upwards force called 'lift' when it is moving. An aeroplane with a mass of 50 000 kg flies level and at a low height. Calculate the total lift force on its wings.

3 A ship takes on 600 000 kg of extra cargo.

Explain why it sinks lower in the water.

4 A racing car tries to set a speed record. At first it gets faster and faster, but soon it reaches a terminal velocity. Draw a diagram and on it mark and label the forces acting on the car at its top speed. Explain which of those forces increases with the speed of the car, and how it sets the terminal velocity.

Key facts:

✔ The forces on an object combine to give an overall resultant force.

✔ A resultant force of zero means all the forces on an object are balanced.

✔ Balanced forces on an object result in steady motion in the same direction, or staying at rest.

✔ Unbalanced forces on an object change its speed or direction of motion.

Check your skills progress:

I can add forces to find the resultant force.

I can explain that terminal velocity is reached when gravity and air resistance balance.

I can explain that an object floats when gravity and upthrust balance.

I can recognise where, and in what direction, push (contact) forces or friction forces will act.

Quick questions

1. What force stops your feet from slipping? [1]

10.15

2. What forces act on a falling leaf? [2]

10.16

3. Using g = 10 N/kg, calculate:

   (a) the weight of an object with a mass of 2 kg [1]

   (b) the mass of an object which weighs 3000 N. [1]

4. Explain why we lubricate machines with oil. [2]

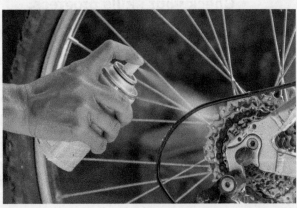

10.17

Connect your understanding

5. Why can a hovercraft move much faster than a normal ship? **[4]**

6. An ice skater turns in a circle at a steady speed. Draw and label a diagram to name *three* forces acting on her skate and mark each with an arrow to show its direction. **[6]**

7. Explain the direction of Earth's gravitational force on different sides of the planet, for example at the north and south poles. Why does nobody fall off? **[4]**

8. A sky-diver, dropping very fast, opens his parachute just a few hundred metres before reaching the ground. Draw and label diagrams and use them to explain the difference between the forces that act:

    (a) when the parachute first opens and

    (b) just before he lands safely. **[6]**

9. Dogs pull a sledge over a flat, icy surface. Their pulling force is 240 N forwards. Air resistance is 160 N and friction from the ice is 60 N.

    (a) Calculate the size and direction of the overall resultant force on the sledge. **[2]**

    (b) Explain how you know whether the sledge is speeding up or slowing down. **[2]**

10. Explain why most of a swimmer's body will normally be under the water level.

10.18 **[3]**

11. Astronauts on a space mission travel from the Earth to the Moon.

10.19

Describe the different sizes and directions of gravitational forces they experience:

(a) close to the Earth [2]

(b) halfway between Earth and the Moon [2]

(c) when they have landed on the Moon. [2]

(d) Is there is any point on the journey when their measured weight would
be zero? Explain where and why. [2]

Challenge questions

12. (a) Explain how you feel your own weight, and why astronauts in a space
station do not feel their weight even though they are still pulled by the
Earth's gravity. [4]

(b) Explain why the astronauts and the space station they are in do not fall
to Earth but stay in orbit. [4]

13. Newton's First Law of Motion predicts that, when there is no force acting on it, an
object moves with a steady speed in a straight line. Explain why it is difficult to
show this in an experiment. What might you try? [5]

1. (a) Malia is playing on a swing.

Select the *two* correct statements.

    a At the highest point Malia has maximum kinetic energy

    b At the lowest point Malia has maximum potential energy

    c At the highest point Malia has no kinetic energy

    d At the lowest point Malia has no potential energy   **[2]**

(b) The drawing shows a solar LED street light. It has a solar panel, a battery and an LED lamp.

    (i) Complete the following sentences:

        The energy source is _____ energy. The battery stores _____ energy. _____ energy is transferred from the battery to the light.   **[3]**

    (ii) For every 10 kJ of energy stored in the battery, 8.5 kJ of useful light energy is output from the lamp. Calculate how much wasted energy is output.   **[1]**

(c) Fossil fuels include coal, oil and natural gas.

    Explain why these fuels are non-renewable.   **[1]**

2. (a) The eight planets in our Solar System are Earth, Jupiter, Mars, Mercury, Neptune, Saturn, Uranus and Venus.

    Name the planet that is closest to the Sun.   **[1]**

(b) The table shows data for three planets: A, B and C.

Planet	A	B	C
Length of day (hours)	9.9	17.2	10.7
Orbit period (Earth days)	4331	30 589	10 747

Write the name of the planet that is furthest from the Sun. [1]

(c)

Write the letter of the picture that shows a solar eclipse. [1]

(d) Choose from these words to complete the sentences below.

12 hours   24 hours   27.3 days   40 days   week   year

The Earth rotates once on its axis every _____. The Moon orbits the Earth

once every _____. The Earth orbits the Sun once every _____. [3]

3. (a) Michael wants to investigate friction. He uses the equipment shown in the drawing. He pulls the block over a table with the newton meter.

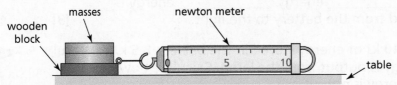

    (i) Predict how the reading on the newton meter changes when he puts more masses on the wooden block. [1]

    (ii) Michael repeats his measurements for each mass three times and works out the average. Explain why. [1]

(b) Two astronauts plan to travel to Mars.

Astronaut	Mass	Weight on Earth	Weight on Mars
Asif	75 kg	750 N	
Aleena		600 N	228 N

   (i) Complete the gaps in the table above. [2]

   (ii) Write the name of the planet with the weakest gravity. [1]

(c) Choose the correct expression to complete the following sentences about a sky diver after she jumps out of plane.

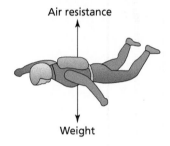

Air resistance

Weight

   (i) At first she falls faster and faster. This is because her weight is _____ the air resistance force.

   a less than

   b greater than

   c the same as

   (ii) After 20 seconds her speed is constant. This is because her weight is _____ the air resistance force.

   a less than

   b greater than

   c the same as [2]

[total 20 marks]

1	2													3	4	5	6	7	0
																			4 **He** helium 2
7 **Li** lithium 3	9 **Be** beryllium 4													11 **B** boron 5	12 **C** carbon 6	14 **N** nitrogen 7	16 **O** oxygen 8	19 **F** fluorine 9	20 **Ne** neon 10
23 **Na** sodium 11	24 **Mg** magnesium 12													27 **Al** aluminium 13	28 **Si** silicon 14	31 **P** phosphorus 15	32 **S** sulfur 16	35.5 **Cl** chlorine 17	40 **Ar** argon 18
39 **K** potassium 19	40 **Ca** calcium 20	45 **Sc** scandium 21	48 **Ti** titanium 22	51 **V** vanadium 23	52 **Cr** chromium 24	55 **Mn** manganese 25	56 **Fe** iron 26	59 **Co** cobalt 27	59 **Ni** nickel 28	63.5 **Cu** copper 29	65 **Zn** zinc 30			70 **Ga** gallium 31	73 **Ge** germanium 32	75 **As** arsenic 33	79 **Se** selenium 34	80 **Br** bromine 35	84 **Kr** krypton 36
85 **Rb** rubidium 37	88 **Sr** strontium 38	89 **Y** yttrium 39	91 **Zr** zirconium 40	93 **Nb** niobium 41	96 **Mo** molybdenum 42	[98] **Tc** technetium 43	101 **Ru** ruthenium 44	103 **Rh** rhodium 45	106 **Pd** palladium 46	108 **Ag** silver 47	112 **Cd** cadmium 48			115 **In** indium 49	119 **Sn** tin 50	122 **Sb** antimony 51	128 **Te** tellurium 52	127 **I** iodine 53	131 **Xe** xenon 54
133 **Cs** caesium 55	137 **Ba** barium 56	139 **La*** lanthanum 57	178 **Hf** hafnium 72	181 **Ta** tantalum 73	184 **W** tungsten 74	186 **Re** rhenium 75	190 **Os** osmium 76	192 **Ir** iridium 77	195 **Pt** platinum 78	197 **Au** gold 79	201 **Hg** mercury 80			204 **Tl** thallium 81	207 **Pb** lead 82	209 **Bi** bismuth 83	**Po** polonium 84	**At** astatine 85	**Rn** radon 86
Fr francium 87	**Ra** radium 88	**Ac**** actinium 89	**Rf** rutherfordium 104	**Db** dubnium 105	**Sg** seaborgium 106	**Bh** bohrium 107	**Hs** hassium 108	**Mt** meitnerium 109	**Ds** darmstadtium 110	**Rg** roentgenium 111									

Key

relative atomic mass
atomic symbol
name
atomic (proton) number

1
H
hydrogen
1

La lanthanoids

Ac actinoids

Elements with atomic numbers 112–116 have been reported but not fully authenticated

Elements 1 to 92 are naturally occurring elements on Earth. Elements 93 and above are man-made.

Glossary

Biology

acid rain: rain that is much more acidic than usual.

adaptation: feature of something that allows it to do a job (function) or allows it to survive.

amphibian: vertebrate with moist skin. It lays jelly-coated eggs in water.

animal kingdom: kingdom that contains organisms that are made of more than one cell and are able to move their bodies from place to place.

antagonistic pair: two muscles that pull a bone in opposite directions.

antiseptic: substance that kills microorganisms but is safe for us to put on our skins.

arachnid: arthropod with eight legs and a body in two sections.

arthropod: invertebrate with jointed legs and a body in sections.

bacterium: type of one-celled organism that is not a plant or animal or fungus. The plural is 'bacteria'.

ball and socket joint: joint where a ball-shaped piece of bone fits into a socket made by other bones.

bar chart: chart that shows data using columns. They are used to compare different sets of things.

biofuel: fuel made using plants or algae.

bird: vertebrate with feathers. It lays eggs with hard shells.

bladder: organ that stores urine.

blood: liquid organ that carries substances around the body.

blood vessels: tube-shaped organs that carry blood around the body.

bone: hard organ that supports or protects the body, or allows movement.

carnivore: animal that eats other animals.

cell: the smallest living part of an organism.

cell membrane: outer layer of a cell that controls what enters and leaves the cell.

cell wall: strong outer covering found in some cells (such as plant cells).

characteristic: feature of an organism.

chloroplast: green part of a cell that makes food using light.

circulatory system: group of organs that get blood around the body.

climate change: changes to weather patterns.

conclusion: decision that you reach. In science, you use evidence from experiments to make conclusions.

conifer: plant with needle-shaped leaves. It produces cones.

consumer: animal that eats other living things.

continuous variation: variation that can have any value within a range.

contract (muscle): when a muscle gets shorter and fatter it contracts.

cytoplasm: watery jelly where the cell makes new substances.

daily change: change in physical factors during the course of a day.

data: numbers and words that can be organised to give information.

decay: when materials break into smaller parts. Microorganisms often cause this.

decomposer: microorganism that causes decay.

deciduous: plant that loses its leaves during a certain season of the year.

deforestation: cutting down forests.

diagnosis: saying what disease someone has.

diaphragm: organ that helps with breathing.

digestive system: group of organs that digest food and get it into the blood.

discontinuous variation: variation that has a distinct set of options or categories.

Glossary

disinfectant: substance that kills microorganisms on surfaces that we touch.

echinoderm: invertebrate with a hard, spiny outer covering.

ecosystem: all the organisms and the physical factors in a habitat.

environment: the other organisms and physical factors around an organism.

evidence: data or observations we use to support or oppose an idea.

excrete: getting rid of wastes made inside an organism.

extinct (life forms): does not exist any more.

eyepiece lens: the lens of a light microscope that you look through.

fern: plant that does not produce flowers or cones but has roots.

fever: high body temperature.

fish: vertebrate with slimy scales. It lays jelly-coated eggs in water.

flower: contains organs used in reproduction (to make seeds).

flowering plant: type of plant that produces flowers.

focusing wheel: wheel on a microscope that you turn to make an image clear.

food chain: list with arrows that shows what eats what in a habitat.

fossil fuels: fuel, such as coal, oil and natural gas, made in the ground over millions of years from dead organisms.

fuel: substance that releases energy.

function: another word for 'job'.

fungus: type of organism that is not a plant or an animal. The plural is 'fungi'.

global warming: increasing temperatures around the Earth and its atmosphere.

greenhouse effect: when gases in the atmosphere trap energy and cause the Earth to warm up.

habitat: the place where an organism lives.

haemoglobin: substance that traps oxygen.

hand lens: another term for magnifying glass.

heart: organ that pumps blood through blood vessels.

herbivore: animal that eats plants.

hibernation: when animals go into a type of sleep during cold seasons.

hinge joint: joint where two bones form a hinge.

hybrid: offspring produced by reproduction between two different species.

infected: when a disease-causing microorganism is in someone, they are infected.

infectious disease: disease that spreads from one organism to another.

insect: arthropod with six legs and a body in three sections.

invertebrate: animal without a skeleton inside it and without a 'backbone'.

joint: place in your skeleton where bones meet.

kidneys: organs that remove wastes from the blood to produce urine.

kingdom: the biggest of the groups that scientists use to classify organisms.

large intestine: organ that absorbs water from undigested food.

leaf: plant organ that makes food for a plant.

life process: something that all living things do.

ligament: cord that attaches bones together.

limewater: clear and colourless liquid that turns milky when carbon dioxide is added.

line graph: graph that shows data points plotted on a grid. Line graphs are often used to show how one thing changes with time. Time is put on the horizontal axis.

liver: organ that makes and destroys substances.

Louis Pasteur: French scientist who discovered that microorganisms spoil food.

lungs: organs that get oxygen into the blood and remove carbon dioxide.

magnification: the amount to which something is magnified.

magnify: to make something appear bigger.

magnifying glass: used to make things appear bigger (magnify them).

mammal: vertebrate with hair. It gives birth to live offspring.

microbe: another word for 'microorganism'.

microorganism: tiny organism. We must use microscopes to see them.

microscope: piece of equipment that magnifies very small things.

migration: when animals move from one area to another as the seasons change.

model: simple way of showing or explaining a complicated object or idea.

mollusc: invertebrate with a large muscle that it uses to move or feed.

moss: plant with small, thin leaves. It does not have roots.

mould: fungus that decays things.

muscle: organ that changes shape. Some muscles move bones.

nervous system: group of organs that control the body.

nocturnal: active at night.

non-renewable: something that will not last forever.

nucleus: control centre of a cell.

nutrient: substance that an organism needs to stay healthy and survive.

nutrition: getting substances needed for survival.

objective lens: the lens in a light microscope that is closest to the specimen. Most light microscopes have several objective lenses, with different magnifications.

offspring: new organism made when parents reproduce.

omnivore: animal that eats both plants and animals.

organ: part of an organism that has an important job (function).

organ system: group of organs working together.

organism: living thing.

ozone depletion: reducing the amounts of ozone.

palisade cell: cell found in plant leaves, which contains many chloroplasts.

physical factor: non-living part of an environment (e.g. wind).

pitfall trap: jar buried in the ground to collect small animals that walk on the ground.

plant kingdom: kingdom that contains organisms that are made of more than one cell and make their own food.

pollutant: substance that causes harm to organisms.

pollution: when organisms are being harmed by a substance.

pooter: device to suck small animals into a collecting jar without harming them.

population: the number of one type of organism in a place.

predator: animal that hunts and eats other animals (called prey).

prediction: what you think will happen in an investigation.

prey: animal that is hunted and eaten by other animals (called predators).

primary consumer: the first consumer in a food chain, which is always a herbivore.

producer: organism that makes its own food, such as a plant.

quadrat: square frame used to take samples in a habitat.

Glossary

range: the highest and lowest values in a set of data.

relax (muscle): when a muscle stops contracting it relaxes.

renewable: something that will not run out.

reproduce: when organisms have young (or offspring).

reptile: vertebrate with dry scales. It lays eggs with a leathery coat.

resource: anything that is needed or used by an organism.

respiration: chemical process that happens in all parts of an organism to release energy.

respiratory system: group of organs that get oxygen into the blood and remove carbon dioxide. Also called the breathing system.

rib: bone that helps to protect your heart and lungs.

root: plant organ that absorbs water from the ground, and holds the plant in place.

root hair cell: plant cell found in roots that is adapted for taking in water quickly.

sample: small portion of something, used to discover what the whole of the thing is like.

scientific method: stages that scientists use to test out their ideas.

scientific question: question that scientists can answer using an experiment.

season: time during the year with a certain set of physical factors.

seasonal change: change in physical factors during the course of a year.

secondary consumer: the second consumer in a food chain, which is always a carnivore.

sensitivity: how an organism detects changes in things inside and around it.

skeletal system: all the bones in your body.

skeleton: another term for your skeletal system.

skin: organ that protects the body and helps it sense things.

skull: collection of bones that protect your brain.

slide: small sheet of glass on which you place a thin specimen.

small intestine: organ that digests food and absorbs it into the blood.

smog: unpleasant chemical fog.

specialised cell: cell with adaptations for a certain job.

species: type of organism. Organisms of the same species reproduce with one another. They have offspring that then have offspring of their own.

specimen: the thing you examine using a microscope.

spore: single cell released into the air by a fungus and which is able to grow into a new fungus.

stage: flat surface on a light microscope where you put a slide.

stem: plant organ that carries substances around a plant.

stomach: organ that helps to digest food.

surface area: the area of a surface, measured in squared units such as square centimetres (cm^2).

symptom: effect of a disease on the body.

tendon: cord that attaches muscles to bones.

tissue: group of cells of the same type.

top predator: the last predator in a food chain.

urine: liquid containing many wastes made inside animals.

vaccine: substance injected into people to stop them getting an infectious disease.

vacuole: storage space inside some cells (such as plant cells).

variable: something that may change.

variation: differences between characteristics.

vertebrae: the bones in your back. The singular is vertebra.

vertebrate: animal with a skeleton inside it, including a 'backbone'.

virus: particle that is only alive when inside a living cell and cannot reproduce.

volume: How much space a substance takes up. Measured in cm³ or litres. Also called 'capacity'.

whole number: number without fractions or a decimal point.

wilt: when a plant droops because it does not have enough water.

yeast: type of fungus with only one cell.

Glossary

Chemistry

absorbent: soaks up liquids.

acid: substance which has a pH of less than 7 on the pH scale.

acidic: having the properties of an acid.

alkali: base that dissolves in water to make a solution with a pH of more than 7.

alkaline: if a base is dissolved in water then the solution is alkaline.

base: substance that neutralises an acid. It has a pH of more than 7 on the pH scale.

boiling: the change of state from liquid to gas.

boiling point: the temperature a substance boils at, and changes from a liquid into a gas.

brittle: breaks when bent.

condensation: the change of state from gas to liquid.

control variable: variable that you keep the same during an investigation.

corrosive: substance that causes burns to the skin and eyes and damages other materials.

crust: the thin outer layer of the Earth.

crystal: solid in which particles are arranged in a regular pattern.

dependent variable: variable you decide to measure in an experiment.

evaporating: the change of state from liquid to gas that happens below the boiling point.

evidence: data or observations we use to support or oppose an idea.

extinct (volcanoes): no longer active.

flexible: can be easily bent and will not break.

fossils: the traces of remains of dead organisms that lived thousands or millions of years ago.

fossil record: collection of fossils identified from different times in the Earth's past that shows how animals and plants have changed over millions of years.

freezing: the change of state from liquid to solid.

hazard symbol: symbol which warns you about the dangers of an object, substance or radiation.

humus: the part of soil which is made from dead or rotting plant material.

indicator: chemical that changes colour in an acid or alkali.

igneous rock: rock formed when magma cools and solidifies.

independent variable: variable you decide to change in an experiment.

inner core: solid layer of the Earth, made of nickel and iron.

lava: magma at the Earth's surface.

litmus: type of indicator which turns red in an acid and blue in an alkali.

magma: molten rock found below the Earth's surface.

malleable: can be formed into different shapes.

mantle: the layer of the Earth beneath the crust. It is mostly solid but it can flow very slowly.

melting: the change of state from solid to liquid.

melting point: the temperature a substance melts at, and changes from a solid into a liquid.

metamorphic rock: rock formed when sedimentary or igneous rocks are changed by very high temperatures and/or pressure.

mineral: solid substance with a fixed chemical composition. Most minerals are crystals.

model: simple way of showing or explaining a complicated object or idea.

neutral: neither acid nor alkali. If soluble, it produces a solution of pH 7.

neutralisation: chemical reaction between an acid and a base which produces a neutral solution.

opaque: light cannot pass through it.

outer core: liquid layer around the inner core of the Earth, made of nickel and iron.

palaeontologist: scientist who studies fossils.

particle theory: model that describes how particles are arranged differently in solids, liquids and gases.

pH scale: scale from 0 to 14 which measures how strong or weak an acid or alkali is.

physical properties: the properties of an object that can be observed and measured.

porosity: the amount of empty space in a material.

prediction: what you think will happen in an investigation.

reliable: measurements are reliable when repeated measurements give results that are very similar.

reversible change: change in a substance that can be changed back again.

scratch test: test to see how hard a rock is, by how easy it is to scratch.

secondary sources: information that has been produced by somebody else.

sedimentary rocks: rocks formed from layers of sediment deposited by water, wind or ice.

sediments: small pieces of rock, such as pebbles, sand and mud.

soil: mixture of small particles of rock, dead animals and plants, water and air.

state of matter: the three forms that a substance can exist in: solid, liquid and gas.

transparent: light can pass through it.

universal indicator: type of indicator which can change into a range of colours depending on whether the solution is acidic or alkaline and how strong it is.

vapour: liquid that has evaporated to form a gas.

variable: something that may change.

volume: How much space a substance takes up. Measured in cm^3 or litres. Also called 'capacity'.

Glossary

Physics

air resistance: the force that acts to slow down an object moving through air. It varies with the size and shape of the object.

axis: imaginary straight line, running through the centre of the Earth between the two poles.

balanced forces: when the resultant force is zero.

chemical energy: energy that can be transferred in a chemical reaction, e.g. burning fuel.

conservation of energy: energy cannot be created or destroyed. The total amount of energy is constant.

eclipse: when one object interferes with our view of another object.

electrical energy: energy that can be transferred from a battery or power supply.

equator: imaginary line running around the circumference of the Earth halfway between the north pole and the south pole.

fossil fuels: fuel, such as coal, oil and natural gas, made in the ground over millions of years from dead organisms.

free fall: the speeding up downwards motion of a body due to its weight.

friction: force between two surfaces that are pressed together. It tries to stop the surfaces sliding over one another.

galaxy: collection of billions of stars held together by gravity. Our Sun with its Solar System is part of the Milky Way galaxy.

gravity: the pull of a large mass like the Earth on other masses near it.

heat: thermal energy that is transferred from a hot object to a colder object.

joule: the scientific unit for energy. Its abbreviation is J. 1000 J = 1 kilojoule (kJ).

kinetic energy: energy stored by an object because it is moving.

light energy: energy that is transferred by a light source.

moon: natural object made of rock or frozen liquid, which orbits a planet.

non-renewable energy resources: resources that cannot be replaced quickly.

northern hemisphere: the half of the Earth's spherical surface that lies north of the equator.

orbit: circular or elliptical path in which one object travels round another. Gravity holds the objects together.

planet: object which orbits around the Sun and is large and heavy enough to have become approximately spherical and to have cleared all other smaller bodies out of its orbit.

pole: the north and south poles mark the ends of the axis about which the Earth spins.

potential energy: the amount of stored energy something has because of its position.

renewable energy resources: resources that can be replaced quickly in natural processes.

resultant force: shows the single total force acting on an object when all the forces acting on it are added up.

scientific law: a theory that has been tested by experiment and shown to be useful and reliable.

seasons: divide the year into four periods of three months each: winter, spring, summer and autumn (or fall). Winter has shortened daylight hours and summer has the longest. On the equator, day and night are always 12 hours each, so there are no seasons there.

Solar System: the Sun and all the other objects that move around it under the control of its gravity.

sound energy: energy that is transferred by a vibrating object making a noise, e.g. a musical instrument.

star: huge mass of gas that is undergoing nuclear reactions. Stars are so hot that they give out light all the time.

terminal velocity: the steady speed of an object with balanced pull and air resistance forces acting on it.

theory: a set of ideas that describes how things work.

thermal energy: energy stored in an object due to its temperature.

tilt: the angle between the Earth's spin axis and the axis of its orbit around the Sun.

upthrust: the upwards force from a liquid on a wholly or partly submerged object.

weight: the size of the pull of gravity for a given mass. You can feel something's weight when you try to stop it from falling.

work: transfer of energy that causes an object to move.

Index

Index

The publishers gratefully acknowledge the permission granted to reproduce the copyright material in this book. Every effort has been made to trace copyright holders and to obtain their permission for the use of copyright material. The publishers will gladly receive any information enabling them to rectify any error or omission at the first opportunity.

p2 NASA/JPL-Caltech/MSSS; p3 Science History Images/Alamy Stock Photo; p4 coxy58/ Shutterstock; p6 Aggie 11/Shutterstock; p7 GIPhotoStock/SCIENCE PHOTO LIBRARY; p9 GUDKOV ANDREY/Shutterstock; p10t Asia Images/Shutterstock; p10b PRABHAS ROY/ Shutterstock; p11 M.A.PUSHPA KUMARA/EPA-EFE/REX/Shutterstock; p14l Alex Mit/ Shutterstock; p14tr koya979/Shutterstock; p14br Puwadol Jaturawutthichai/Shutterstock; p16l Chung Sung-Jun/Getty Images; p16r Chung Sung-Jun/Getty Images; p17 Jarva Jar/ Shutterstock; p19 WIS Bernard/Getty Images; p23 R.W. HORNE / BIOPHOTO ASSOCIATES/ SCIENCE PHOTO LIBRARY; p25 Triff/Shutterstock; p27l Jose Luis Calvo/Shutterstock; p27r Rattiya Thongdumhyu/Shutterstock; p28 STEVE GSCHMEISSNER/SCIENCE PHOTO LIBRARY; p30 Leonardo da/Shutterstock; p36 POWER AND SYRED/SCIENCE PHOTO LIBRARY; p38l Villiers Steyn/Shutterstock; p38c Natalia van D/Shutterstock; p38b POWER AND SYRED/ SCIENCE PHOTO LIBRARY; p39t Dennis Krunkel Microscopy/Science Photo Library; p39b Dennis Krunkel Microscopy/Science Photo Library; p41 Smith Collection/Gado/Getty Images; p43 BSIP SA/Alamy Stock Photo; p44 vm2002/Shutterstock; p46t SCIENCE PHOTO LIBRARY; p46c riopatuca/Shutterstock; p46b Satirus/Shutterstock; p47 US National Library of Medicine/ Science Photo Library; p49 Malcolm Fairman/Alamy Stock Photo; p50 Geography Photos/ Getty Images; p53 Rattiya Thongdumhyu/Shutterstock; p55 reptilesforall/Shutterstock; p56 Zoonar GmbH/Alamy Stock Photo; p57tl Nigel Cattlin/Alamy Stock Photo; p57tc SCIENCE PHOTO LIBRARY; p57tr Andy Harmer/Science Photo Library; p57bl Philippe Psaila/Science Photo Library; p57bc Jack Barr/Alamy Stock Photo; p57br Pascal Goetchluck/Science Photo Library; p59t Patagonian Stock AE/Shutterstock; p59tc BLUR LIFE 1975/Shutterstock; p59bc The Natural History Museum/Alamy Stock Photo; p59b Cosmin Manci/Shutterstock; p60l kwanchai.c/Shutterstock; p60r Stephane Didouze/Shutterstock; p61t Christopher Wood/ Shutterstock; p61b robynleigh/Shutterstock; p63r David Havel/Shutterstock; p63l dani92026/ Shutterstock; p66l Studio-Neosiam/Shutterstock; p66c Anan Kaewkhammul/Shutterstock; p66r Eric Isselee/Shutterstock; p70t Salienko Evgenii/Shutterstock; p70b Dr Morely Read/ Shutterstock; p71 NASA's Goddard Space Flight Center; p72 Martyn Jandula/Shutterstock; p73t Anticiclo/Shutterstock; p73b ERIC CABANIS/Getty Images; p74 Sim Creative Art; p80 gabriel12/Shutterstock; p81 Rawpixel.com/Shutterstock; p82 nattanan/Shutterstock; p83t JONATHAN PLEDGER/Shutterstock; p83c Giusparta/Shutterstock; p83bl haningpixels/ Shutterstock; p83bc ©Royal BC Museum; p83br Donvan van Staden/Shutterstock; p84 Brenda Smith DVM/Shutterstock; p88 row1, l Kletr/Shutterstock; p88, row 1, cl Anton Kozyrev/Shutterstock; p88, row 1, c Portogas D Ace/Shutterstock; p88, row 1, cr Vitalij Terescsuk/Shutterstock; p88, row 1, r Jiri Vaclavek/Shutterstock; p88, row 2, l Eric Isselee/ Shutterstock; p88, row 2, cl Troscha/Shutterstock; p88, row 2, cr goir/Shutterstock; p88, row 2, r Scott Sanders/Shutterstock; p89l Eric Isselee/Shutterstock; p89cl PetlinDmitry/ Shutterstock; p89c Andrew Burgess/Shutterstock; p89cr Tsekhmister/Shutterstock; p89r cyo bo/Shutterstock; p90tl By Kletr/Shutterstock; p90tr goir/Shutterstock; p90bl Eric Isselee/ Shutterstock; p90br Anton Kozyrev/Shutterstock; p91l Filipe B. Varela/Shutterstock; p91cl Gita Kulinitch Studio/Shutterstock; p91cr Johannes Kornelius/Shutterstock; p91r NIPAPORN PANYACHAROEN/Shutterstock; p92 L. Lee Grismer; p94t Don Mammoser/Shutterstock; p94c Joule Sorubo/Shutterstock; p94b ESB Professional/Shutterstock; p95 Red Images, LLC/

Alamy Stock Photo; p100l Aumsama/Shutterstock; p100r Daniel Prudek/Shutterstock; p101 Ecopint/Shutterstock; p109 Sophie James/Shutterstock; p110r Honza Krej/Shutterstock; p110l Duplass/Shutterstock; p114 showcake/Shutterstock; p117l Steven Coling/Shutterstock; p118t imamchits/Shutterstock; p118c Mirek Nowaczyk/Shutterstock; p118b Martin Karius/ Alamy Stock Photo; p119t Nor Gal/Shutterstock; p119bl Leon Chong/Shutterstock; p119bc Chutima Chaochaiya/Shutterstock; p119br Cinematographer/Shutterstock; p120 Thanamat Somwan/Shutterstock; p121t Tupungato/Shutterstock; p121b B-C Images/Getty Images; p122t Jiri Vaclavek/Shutterstock; p122b e_rik/Shutterstock; p127 V J Matthew/Shutterstock; p128 Artography/Shutterstock; p129tl Tom Grundy/Alamy; p129tc Alexlukin/Shutterstock; p129tr vvoe/Shutterstock; p129cl Folkin Oleg/Shutterstock; p129cc Tyler Boyes/Shutterstock; p129cr Branko Jovanovic/Shutterstock; p129b urbazon/Shutterstock; p130 beboy/Shutterstock; p131t Bylikeova Oksana/Shuttestock; p131b Dario Lo Presti/Shutterstock; p132tr SAPhotog/ Shutterstock; p132cr Bjoern Wylezich/Shutterstock; p132cl Michael Dunning/Getty Images; p132br Walter Bilotta/Shutterstock; p133t Tyler Boyes/Shutterstock; p133b Antonov Roman/ Shutterstock; p135 em faies/Shutterstock; p136t Tatsiana Salayuova/Shutterstock; p136b forstbreath/Shutterstock; p138t Sabena Jane Blackbird/Alamy Stock Photo; p138b Juergen Ritterbach/Alamy Stock Photo; p139 Alice-Photo/Shutterstock; p141t Galena Andruskho/ Shutterstock; p141b dieKleinert / Alamy Stock Photo; p144 5 Second Studio/Shutterstock; p146 ACORN 1/Alamy Stock Photo; p147tr Art Directors & Trip/Alamy Stock Photo; pp147l Inna Bigun/Shutterstock; p147br Shawn Hempel/Shutterstock; p149 Asian Images/ Shutterstock; p151 NaMaKuKi/Shutterstock; p154 irin-k/Shutterstock; p159 sirtravelalot/ Shutterstock; p160 CG Stocker/Shutterstock; p161l Shahril KHMD/Shutterstock; p161r bunnyphoto/Shutterstock; p162 PRILL/Shutterstock; p163 Seksan 99/Shutterstock; p167 Elina-Lava/Shutterstock; p169 Charles Wollertz/Alamy Stock Photo; p170 Alex Treadway/ Getty Images; p171 Forrest Anderson/Getty Images; p175 NASA/Apollo Imagery; p176l Suriya Desatit/Shutterstock; p176c paula french/Shutterstock; p176r gil80/Shutterstock; p177r John Henshall/Alamy Stock Photo; p177l kak2s/Shutterstock; p179 norph/Shutterstock; p181 Ian McKinnell/Alamy Stock Photo; p183l NASA/Aubrey Gemignani; p183r NASA Ames Research Center/Brian Day; p184t CARLOS CLARIVAN/SCIENCE PHOTO LIBRARY; p184b MIKKEL JUUL JENSEN/SCIENCE PHOTO LIBRARY; p185 KUCO/Shutterstock; p186l NASA; p186r NASA/JPL/ STScl Hubble Deep Field Team; p189 Zoonar GmbH/Alamy Stock Photo; p191t LuckyImages/ Shutterstock; p191b Inge Schepers Sports and Events Photography/Alamy Stock Photo; p192 Robert McLean/Alamy Stock Photo; p195 NASA/SCIENCE PHOTO LIBRARY; p198 Sky Antonio/ Shutterstock; p200t Halfpoint/Shutterstock; p200c kuzmafoto/Shutterstock; p200b senee sriyota/Shutterstock; p201 Jacob Lund/Shutterstock; p202 Arvind Balaraman/Shutterstock; p204l NASA Ames Research Center/Brian Day; p204r NASA/Aubrey Gemignani.